IET CONTROL ENGINEERING SERIES 75

Advanced Control for Constrained Processes and Systems

Other volumes in this series:

Advanced Control for Constrained Processes and Systems

Fabricio Garelli, Ricardo J. Mantz and Hernán De Battista

The Institution of Engineering and Technology

Published by The Institution of Engineering and Technology, London, United Kingdom

The Institution of Engineering and Technology is registered as a Charity in England & Wales (no. 211014) and Scotland (no. SC038698).

The Institution of Engineering and Technology
Michael Faraday House
Six Hills Way, Stevenage
Herts, SG1 2AY, United Kingdom

www.theiet.org

British Library Cataloguing in Publication Data
A catalogue record for this product is available from the British Library

ISBN 978-1-84919-261-3 (hardback)
ISBN 978-1-84919-262-0 (PDF)

Typeset in India by MPS Ltd, a Macmillan Company
Printed in the UK by CPI Antony Rowe, Chippenham, Wiltshire

To Lau (F.G.), Lilian (R.M.) and Vale (H.D.B.)

Contents

Chapter 1
An introduction to constrained control

1.1 Motivations

In every real control loop, there exist physical limits that affect the achievable closed-loop performance. Particularly, it is well known that mechanical stops or technological actuator limitations give rise to unavoidable constraints at the input to the plant, which must be taken into account to meet performance specifications or safety operation modes.

In single-input single-output (SISO) systems, these physical limits at the plant input are the principal cause of highly studied problems like controller and plant windup. However, there are other types of constraints that have been dealt with to a much lesser extent but that also have an effect on the achievable performance. For instance, the few degrees of freedom of industrial controllers (typically PID) or plants having non-minimum phase features will generally restrict the evolution of the controlled variables. What is more, some well-known problems such as bumpy transfers can be attributed to the fact of constraining the controller types and their switch scheduling. Thus, there also exist structural or dynamic constraints that together with performance specifications, environmental regulations or safety rules usually require system states or outputs to be bounded.

Although constrained control problems have been studied primarily in SISO systems, the majority of the *real-world* processes have more than one variable to be controlled and possess more than one control action for this objective. These systems are called multivariable systems or multiple-input multiple-output (MIMO) systems. Actually, SISO systems frequently are a given subsystem of an overall MIMO system.

Multivariable systems can be found almost everywhere. In the bathroom of a house, the water temperature and flow rate are important variables for a pleasant shower. In chemical processes, it is commonly required to simultaneously control pressure and temperature at several points of a reactor. An automated greenhouse should ensure that the lighting, relative humidity and temperature are adequate for a given cultivation. A robot manipulator needs six degrees of freedom to have a full positioning rank, whereas in a plane or a satellite there are dozens of variables to be controlled.

There are some phenomena that are present only in MIMO systems and do not occur in SISO systems. For example, the presence of *directions* associated to input/output vectors is exclusive to MIMO systems. On this account, a multi-variable system may have a pole and a zero at the same location that do not cancel each other, or in a minimum-phase MIMO system the individual elements of the transfer matrix might have their zeros in the right-half plane (RHP), or vice versa. Nevertheless, the most distinctive property of a multivariable system is probably the crossed coupling or *interactions* between its variables. In effect, in a MIMO system each input variable affects not only its corresponding output but also all the remaining controlled variables of the system. This makes controller design a difficult task, and in most applications precludes one from doing it as if the system consisted of multiple mono-variable loops, since the gains of a single-loop controller will have impact on the other loops and may even cause instability. This is the reason why crossed interactions are generally considered the main difficulty of multivariable control systems [144]. Therefore, in a multivariable process the effects and demands of the aforementioned system constraints are worsened because of the *directionality* and *interactions* present in this kind of plants. The search for solutions to such problems has motivated several research works in the previous years [46,63,123–125,127].

Despite the large number of existing methods to address constrained control problems, a common practice of engineers is to design control systems using conventional methods in such a way that they avoid reaching the system limits and, at the same time, achieve a reasonable performance for a given operating region. However, this conservative approach is seldom feasible in complex systems control or high-performance applications. Moreover, even in relatively simple industrial problems, the resulting closed-loop performance can be significantly improved if system constraints are taken into account.

Contrary to what some practising engineers believe, this does not necessarily require a complete redesign of the control system and abandoning their valuable experience on nominal control design. Indeed, the basic ideas behind the simplest anti-windup (AW) schemes highly accepted in industry can be further exploited to gain robustness, closed-loop performance and design simplicity in more complex control problems with different kinds of constraints, while preserving conventional control methods for the nominal controller design. This will be a major topic in the book: the first part (Chapters 1–3) mainly devoted to SISO constrained systems and the second part (Chapters 4–8) devoted to some relevant multivariable control problems.

1.2 Types of constraints

Let us start by giving a classification of system constraints affecting closed-loop performance. As could be noted from the introductory comments, constraints shall be understood in a 'wide sense'. That is, we will refer to system constraints not only to mention physical limits of the control loop components but also to mention any other structural constraint or dynamic restriction that affects closed-loop performance and can be tackled by delimiting a given signal in the loop.

Basically, we broadly classify the constraints from their source type into three categories:

- Physical limits
- Structural constraints
- Dynamic restrictions

Any of these can be responsible for performance degradation and may consequently generate the necessity of bounding input, output or internal variables of the process under control (see Figure 1.1).

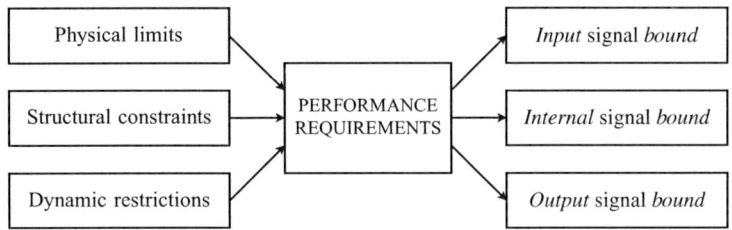

Figure 1.1   Types of constraints and their consequent signal bounds requirements

1.2.1   Physical limits

These are directly related to physical or technological limitations of the elements comprising the control loop. This is the most common type of constraint in the sense that this is what engineers generally refer to when talking about constrained systems.

Some examples, although obvious, seem to be opportune:

- Every electrical engine has a voltage limit and a maximum speed that should not be exceeded.
- A valve can be opened neither more than 100% nor less than 0%.
- The slew rate of a hydraulic actuator will always be limited.

In closed-loop systems, performance requirements frequently lead the control action to hit these physical limitations at the plant input, i.e. actuator saturation is reached. Saturation can occur in either amplitude, rate of change or higher-order signal derivatives, such as acceleration. If the control action exceeds these limits for any reason, severely detrimental behaviours may occur. Therefore, the control system design must somehow account for the unavoidable actuator limits by confining the signal at the plant input to adequate values.

However, although physical limits generally appear at the plant input, they do not exclusively require delimiting the commanded signal to the plant. In effect, we will see later that this kind of constraint may also demand restricting internal or output system variables to avoid performance degradation.

In the next section, we briefly present some of the most common effects produced by this kind of constraints. To emphasise its significance, it is interesting to recall that physical limits, frequently present in simple industrial control problems, have also been involved in extremely serious accidents, like various aircraft crashes and environmental disasters [128].

1.2.2   Structural constraints

We include here the restrictions that are associated with structural limitations of either the plant or the controller. Some examples are as follows:

- Plants in which pre-existing automation devices and circuitry update are not feasible, but new specifications should be met.
- Closed-loop systems with a simple and fixed controller structure, such as P or PI controllers.
- Linear controllers coping with non-linear systems.
- Multiloop or decentralised control structures in multivariable applications.

Although we will deal with many other examples of structural constraints throughout the book, PID controllers are probably the most illustrative case. Not only are they highly accepted in industry applications, but the use of any other type of controller is often turned down. Despite their well-known advantages, it is also true that they are not able to deal with every type of process and specification. Thus, if the controller type cannot be changed, it will result in an additional constraint for the control designer when trying to achieve quite demanding control objectives.

1.2.3   Dynamic restrictions

With this term, we refer to those dynamic characteristics of the process to be controlled (or the controller to be employed) that directly affect the achievable closed-loop performance and the evolution of the signals in the control loop. For instance,

- Plants with RHP zeros and poles.
- Systems or controllers with particular combinations of pole and zero locations.
- Non-linear processes in which detrimental dynamic behaviours are excited by the growth of a given internal variable.

It is well known that RHP zeros produce undesired inverse responses in the controlled variable. Also, systems with a stable zero closer to the origin than the dominant poles may give rise to large overshoots in the step response. In complex processes with non-linear behaviours, the regulation of a variable of interest can lead auxiliary variables to dangerous or undesirable regions (see the last case study in Chapter 2).

Consequently, this group of constraints may also translate into bounding requirements on the loop signals because of performance specifications or safety operation. This will be pointed out in Section 1.4 and further addressed from Chapter 3.

1.3 Some typical effects of constraints

This section discusses some classic examples of the undesired effects that constraints may have on the closed-loop responses. We focus now on the effects related to physical limits at the plant input. Other problems arising from the different types of constraints previously mentioned are briefly enounced in this chapter (see the next section), but they will be described and illustrated afterwards throughout the remainder of the book.

1.3.1 Controller windup

As mentioned earlier, the closed-loop bandwidth requirements often lead the control signal to reach the physical actuator limits at the plant input. When this occurs, unless an action is taken to avoid it, the feedback loop is broken and the system operates in an open-loop fashion. Certainly, once the limit is reached, the actuator continues giving its maximal (or minimal) output independently of the controller output, until it returns to its 'allowed region'. This brings about an overreaction of the controller, which in turn produces a significant degradation of the closed-loop response with respect to the one expected for actuator linear operation. The resulting response typically presents large overshoots and/or long settling times.

Since the effect was originally attributed to the fact that the controller integrator winds up due to input saturation, it was called *integrator windup*. However, it was then observed that the same effects could actually be caused by any other unstable or sufficiently slow controller dynamics, and so the problem is generally known as *controller windup*. There exists a vast bibliography devoted to the study of controller windup and its possible solutions, among which some of the most cited works are References 5, 55, 72 and 96.

To illustrate this phenomenon, the following example is presented:

Example 1.1: Consider a simple model corresponding to a distillation column with a PI controller and a Smith predictor as shown in Figure 1.2. The plant model is given by

$$P(s) : G_p(s)e^{-l_p s} = \frac{0.57}{(8.60s + 1)^2} e^{-18.70s} \tag{1.1}$$

whereas the Smith predictor and the PI controller, designed by means of relay auto-tuning techniques in Reference 53, are

$$\tilde{P}(s) : G_m(s)e^{-L_m s} = \frac{0.57}{(7.99s + 1)^2} e^{-18.80s} \tag{1.2}$$

$$PI(s) : K_p \left(1 + \frac{1}{sT_i} \right) = 1.75 \left(1 + \frac{1}{7.99s} \right) \tag{1.3}$$

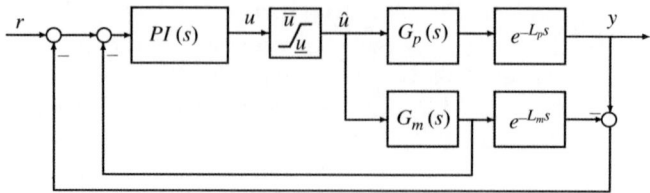

Figure 1.2   PI control and Smith predictor for plant (see (1.1)) with input saturation

The objective here is not to evaluate the behaviour of the set PI–Smith predictor, since any other controller could have been considered for actuator linear operation. Instead, we aim at showing how the closed-loop performance is deteriorated because of input non-linearities, even when modern design techniques are employed.

The simulation results for the closed-loop system without actuator constraints are plotted with dotted line in Figure 1.3. The controlled variable response with an overshoot of approximately 15% and a settling time (5%) of 50 [time units] is then considered the desired response for this control system. In the lower box, the controller output evolution (u, also in dotted line) can be observed, which coincides with the plant input ($\hat{u}$) for each time instant.

The solid curves in Figure 1.3 correspond to the closed-loop response when actuator constraints $\hat{u} \leq \bar{u}$ are considered, with $\bar{u} = 2.2$, $\bar{u} = 2$ and $\bar{u} = 1.9$. As the lower part of the figure makes evident, the controller output (dashed lines) exceeds

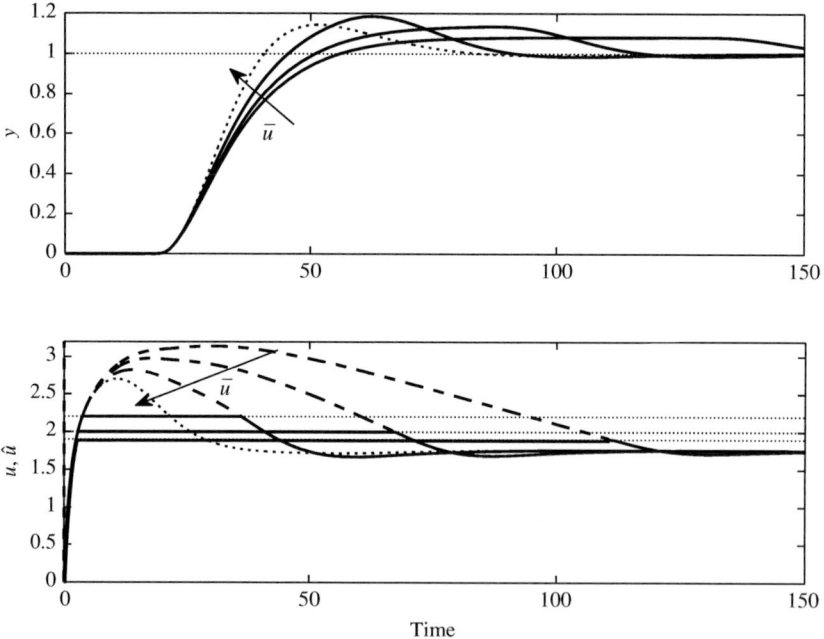

Figure 1.3   Controller windup for different saturation values

the available control amplitude during long periods of time. This leads the output y to evolve slower than in the unconstrained case, and therefore the error e poorly decays. For this reason, the PI integral term winds up, taking a value much greater than when the actuator operates in its linear region. Hence, even when the output y gets close to the reference r, the control action u continues to be saturated because of the integral term contribution. This is a typical controller windup effect, whose principal consequence is in this case the long settling time. Figure 1.3 shows how the settling time increases as saturation becomes harder (24% of increment for $\bar{u} = 2.2$, 73% for $\bar{u} = 2$ and 250% for $\bar{u} = 1.9$).

As we will see in Chapter 2, the undesired behaviour caused by controller windup can be avoided by properly bounding the commanded signal to the plant input, so as to prevent actuator saturation. Then, controller windup provides an example of physical limits requiring input signal delimitation to attain performance demands (see Figure 1.1).

1.3.2   Plant windup

Not only slow/unstable compensator modes or integral actions may produce undesired transients when combined with physical input constraints, but a similar effect to controller windup can also be observed in closed-loop systems with static controllers such as P controllers or state feedback configurations. This is obviously not attributable to the controller dynamics, but it is related to the dynamics of the process under control, since plant states cannot be brought to their steady-state values fast enough due to input saturation. Consequently, although less studied and formally recognised, this effect is known as *plant windup* [60].

We next present an example to demonstrate the problem of plant windup when using a proportional controller.

Example 1.2: Let the third-order system

$$P(s) = \frac{s^2 + 15s + 60}{s^3 + 3s^2 + 2s + 1} \tag{1.4}$$

be controlled by a proportional controller with gain $k_p = 100$, and consider a prefiltering of the reference carried out via

$$F(s) = \frac{10.002}{s + 10} \tag{1.5}$$

From the state-space representation of the system

$$\dot{x} = \begin{bmatrix} 0 & 1 & 0 \\ 0 & 0 & 1 \\ -1 & -2 & -3 \end{bmatrix} x + \begin{bmatrix} 0 \\ 0 \\ 1 \end{bmatrix} u \tag{1.6}$$

$$y = \begin{bmatrix} 60 & 15 & 1 \end{bmatrix} x$$

we can obtain the output and state evolution when a unitary step reference is applied. The results are plotted as dotted lines in Figure 1.4. As can be appreciated, the response exhibits a linear behaviour with a vanishing tracking error.

If we now consider a physical limit (saturation) such that the plant input is constrained to $|\hat{u}| \leq \bar{u} = 1$, the transient response is seriously deteriorated, presenting both long-settling times and large overshoots (solid lines in Figure 1.4). These effects cannot be associated with a controller windup phenomenon since the controller has no dynamic modes. Here, input saturation winds up the system states to values from where they are not able to come back fast enough due to the constrained input signal, thus affecting the output signal. See particularly how the decay rate of the first state x_3 is limited by input saturation.

These undesired effects can even lead to oscillatory responses for greater reference changes or harder physical limits. In general, whenever the closed-loop dynamics is excessively fast with respect to the input limits, there is risk of plant windup.

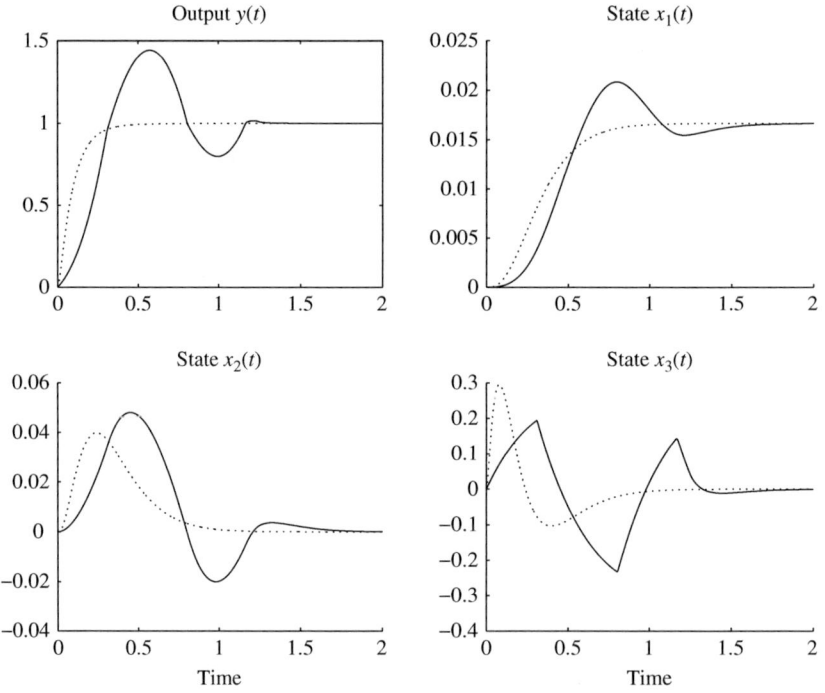

Figure 1.4   *Plant windup effect on output and state responses*

Differing from controller windup, to overcome plant windup it may be necessary to delimit an internal plant state, as will be seen in the next chapter. Then, remembering Figure 1.1, plant windup provides an example of performance degradation caused by physical limits, which can be tackled by delimiting an internal signal.

1.3.3   Control directionality problem

As was claimed in Section 1.1, when designing multivariable systems one has to deal with interactions among the loops. To reduce or avoid such cross-coupling, the plant input vector must have a given direction, determined by a centralised MIMO controller. However, if individual constraints affect each of the controller outputs independently, the plant input direction required for closed-loop decoupling can be drastically altered, for instance, because of the saturation of only one vector component. This effect, also caused by physical input constraints, is known as *control directionality* problem and usually produces strong closed-loop couplings apart from other performance deterioration.

We briefly introduce this problem here with a simple example, skipping the detailed study of multivariable system decoupling and control directionality, since it will be one of the main subjects of Chapters 4 and 5.

Example 1.3: Consider a two-input two-output plant with the following nominal model [49]:

$$P(s) = \frac{1}{(s+1)^2} \begin{bmatrix} s+2 & -3 \\ -2 & 1 \end{bmatrix} \tag{1.7}$$

We assume that the following decoupled transfer matrix is aimed for the closed-loop system:

$$T(s) = \frac{-9(s-4)}{4(s^2 + 4s + 9)} \begin{bmatrix} 1 & 0 \\ 0 & \dfrac{10}{s+10} \end{bmatrix} \tag{1.8}$$

This is accomplished with the centralised controller

$$C(s) = \frac{-9(s+1)^2}{4s} \begin{bmatrix} \dfrac{1}{s+6.25} & \dfrac{30}{s^2 + 14s + 71.5} \\ \dfrac{2}{s+6.25} & \dfrac{10(s+2)}{s^2 + 14s + 71.5} \end{bmatrix} \tag{1.9}$$

which can be synthesised following the ideas described in Chapter 4.

Assuming ideal actuators, a linear decoupled response is obtained when the system is excited with a negative reference change in the second channel (shown with dotted lines in the left graphic in Figure 1.5).

Introducing now two identical saturating actuators whose linear operations are confined to the range $[\underline{u}_i, \overline{u}_i] = [-2, 2]$, $i = 1, 2$, the achieved performance is significantly worsened (solid lines in Figure 1.5). In effect, the closed-loop decoupling for which the controller had been designed is completely lost. The causes of this interaction can be found at the bottom left of the figure, where it is observed how the controller outputs u_1 and u_2 (dashed lines) differ from the plant inputs $\hat{u}_1$ and $\hat{u}_2$

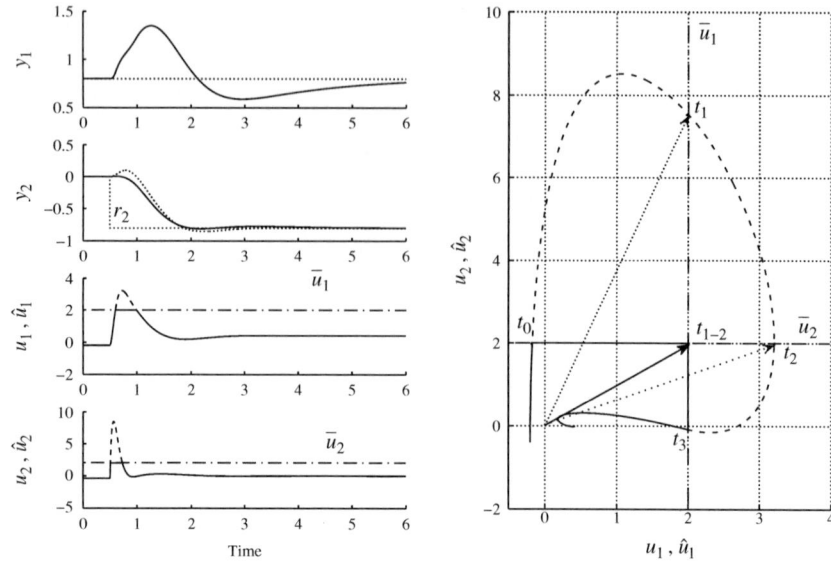

Figure 1.5   Response of decoupled NMP system to a change in r_2, with (solid and dashed lines) and without (dotted lines) input saturation

(solid lines) as a consequence of saturation. As the right box in Figure 1.5 shows, this leads to a transient directionality change of the plant input with respect to the controller output (between t_0 and t_3, $t_0 < t_1 < t_2 < t_3$). Particularly, it can be seen how the controller output direction changes from $t_1 = 0.61$ [time units] to $t_2 = 0.74$ [time units] (between the two longest arrows), while the plant input direction remains unchanged during this period (shortest arrow), when both $\hat{u}_1$ and $\hat{u}_2$ are simultaneously saturating.

1.4   Other constraint implications

Apart from the undesired effects of physical limits described beforehand, there are many other troubles in control system design that may require delimiting the excursion of signals in the loop (recall Figure 1.1). Although most of these problems are not usually addressed by conventional constrained control literature, they are closely related to the remaining types of restrictions defined in Section 1.2, namely structural or dynamic constraints. As we have already mentioned, we will here limit to enounce some of the main problems to be coped with, whereas they will be discussed and described in greater detail in Chapter 3 and at the beginning of Chapters 5–8.

For instance, the output overshoot of a second-order system controlled by a P controller will increase as the desired closed-loop bandwidth does. Thus, an output constraint may be broken if fast responses are demanded and the controller structure cannot be modified. More generally, a similar situation can be found for every linear controller (typically PI or PID industrial controllers) designed for a given operating point but manipulating a real plant, which is actually non-linear and uncertain, as closed-loop requirements are tightened up.

Another typical phenomenon that can be associated with a structural constraint is the *bumpy transfer*, i.e. the jump at the plant input caused by manual–automatic or controller switching, which deteriorates the control system response. In effect, such mode switching can be associated with the fact of using a single linear controller for each operating point of a non-linear process and a trivial scheduling strategy. If no other switching policy (like linear parameter-varying control) is welcomed and non-linear controllers are not accepted, the suppression of the jumps at the plant inputs and their associated effects must be performed by limiting the controller states or input signals. This is usually referred to as *bumpless transfer* and will be discussed in Chapter 8.

Some other signal-bound requirements arising from structural constraints that we will cover in the book are as follows:

- Confining the controlled variable to a given range of values under pre-existing control structures, which cannot be altered, in presence of external disturbances.
- Matching the output of a given plant with the input of a serial connected process or device.
- Limiting the crossed coupling in MIMO systems when the controller is composed of individual SISO controllers for each loop (multiloop control).

Furthermore, the following constraints caused by dynamic restrictions of the plant will be tackled to meet performance specifications:

- To maintain an internal or alternative output variable, which affects the controlled variable of a non-linear system, running along its operational limits.
- To avoid inverse responses in the main controlled variable of non-minimum phase MIMO processes while delimiting the interaction on the remaining variables.

Regarding the group of constraints listed just above, it is important to remark that they are strongly related to the so-called fundamental closed-loop design limitations, for which there are in the literature several analytical measures and tools, e.g. the Bode and Poisson integral constraints.[1] However, as was mentioned earlier, we will focus on fulfilling the signal bounds that result from these unavoidable performance constraints. That is, an engineering approach will be adopted in which explicit and direct signal constraints can be determined by the control designer to accomplish a given goal (see Section 1.6).

---

[1] Even though this subject will be revisited, for further details on fundamental design limitations the reader is referred to References 67 and 110 and references therein.

1.5   Different approaches to constrained control

This book is not intended to give a complete overview of all the existing control methods to deal with constraints. Interesting books related to constrained control are References 46, 51, 60, 62, 69, 107, 121 and 156. Basically, the existing techniques to address constrained control problems can be divided into two categories:

- *One-step* approaches, primarily optimisation techniques.
- *Two-step* approaches, like the well-known AW techniques.

Within the first group, the main avenues are based on either optimal control theory or numerical optimisation. This latter approach includes model predictive control (MPC), a very popular discrete technique for the control of chemical processes. At each sampling time, the implicit form of this algorithm solves an online optimisation problem to compute the control inputs over a future time horizon, making use of a process model to predict the future response. Then, only the first computed control value over the horizon is actually applied to the process. At the next sampling time, the optimisation problem is updated by shifting the horizon forward by one time step (the so-called receding horizon) and solved using the new measures from the plant.

MPC is a *one-step* procedure because the online optimisation can incorporate various constraints, including control or quality objectives, 'from scratch'. Indeed, the main advantage of MPC is probably its ability to deal with either input or output constraints so that nominal performance specifications are met. It is also argued that this is the reason why a significant portion of the literature is devoted to it. However, MPC has often been criticised for being applicable only to relatively slow processes due to the online program running time. Moreover, although it is numerically very powerful, its analytical treatment is quite involved. Since this approach will not be followed here, the reader is referred to Reference 102 for a tutorial about MPC and to Reference 51 for a more recent analytical treatment of these techniques.

The *two-step* approaches (also called evolutionary approaches) are the approaches in which the nominal controller is designed without explicitly considering constraints (first step), and then a compensation loop is added to reduce the adverse effects of constraints as much as possible. The principal motivation for the development of this type of approaches is that they can be intuitively understood, designed and tuned. In fact, any desired control strategy, including those conventional techniques for which operators are usually more trained, can be utilised at the first step for the design of the main control loop. This is probably the main reason why two-step techniques constitute the most widely spread approach to constraints in industrial control.

All the so-called AW techniques rely on this design procedure. Among them, we can distinguish two groups of methods. The first group includes the conventional anti-(reset) windup methods, which modify the controller (usually its integral term) as a function of the plant input or its difference with the controller output.

See, for instance, the methods described in References 4, 5 and 16. The second group includes the algorithms conditioning the reference signal so that the 'realisable reference' is sought, which is, by definition, the reference that if applied from the beginning would just avoid the differences between the controller output and the plant input, i.e. saturation. This highly successful approach was originally formulated by Hanus and co-workers [54,55], and was afterwards generalised by Walgama *et al.* [141]. Despite the vast effort to unify and formalise AW techniques [60,72,96], most of the two-step algorithms are basically *ad hoc* methods without the background of a well-established theory. Also, since they are mainly conceived as AW schemes, two-step approaches generally cope with input constraints only.

1.6   Book philosophy

This book is aimed at providing a practical and unified approach to deal with several control problems caused by either physical, structural or dynamic constraints. To this end, we adopt a two-step approach in an attempt to take advantage of its intuitive features, while at the same time we explore the ways to further extend its applicability and to systematise its design.

Particularly, we introduce a two-step algorithm, which can handle constraints on both the manipulated (input) and the controlled (output) variables, as well as internal state limitations. The resulting control strategies are based on reference-conditioning ideas, implemented by means of an auxiliary or supervisory loop, which employs a discontinuous action to generate the maximum reference signal compatible with the system constraints. Although design simplicity is a book priority, well-established Variable Structure Systems (VSS) theory and sliding mode (SM) related concepts are used for the methodology analysis. This theoretical approach not only gives the proposals a rigorous though conceptually simple mathematical support, but also provides them with distinctive robustness features and reduced dynamic behaviours. To encourage the method implementation, SISO techniques are exploited as much as possible, even when dealing with MIMO control problems. Then, a multivariable sliding mode methodology is also introduced to allow further developments.

The approach taken is *unified* because the same basic ideas allow tackling many constrained control problems (those with different types of constraints, involving either linear or non-linear models, SISO or MIMO processes, etc.), and *practical*, since the resulting techniques are very easy to be implemented and complementary to any other control strategy that has been used in the main control loop.

As was mentioned in the previous section, the book does not intend to cover all the existing control methods to deal with constraints. Instead, the book focuses its attention on the aforementioned unified approach conceived to cope with a wide variety of control problems. It should not be expected either a purely mathematical treatment of the topic, as is quite usual in constrained control literature. On the contrary, an engineering perspective combining rigorous analysis with focus on concrete applications is the main approach.

1.7   Short outline of the main problems to be addressed

The first part of the book (Chapters 1–3) is devoted to SISO constrained control systems. After introducing in this chapter the book motivations together with a brief discussion of the topic, the main approach of the book is presented in Chapter 2. The basic idea is first described and analysed in light of VSS theory. Then, the methodology is extended to deal with signal bounds at any part of the control systems, arising from either physical, structural or dynamic constraints and performance requirements. The proposed technique is finally cast within the framework of non-linear systems and geometric invariance concepts, while its distinctive and robustness features are also shown.

To illustrate the practical potentials of the approach, Chapter 3 presents several case studies with different constraints to be fulfilled: (1) the pitch control of wind turbines with both amplitude and rate actuator saturation; (2) a clean-hydrogen production plant in which the electrolyser specifications require output bounds; (3) the tracking speed autoregulation of robotic manipulators to avoid path deviations; and (4) the regulation of ethanol concentration below a given threshold in the fed-batch fermentation of an industrial strain for overflow metabolism avoidance.

The second part of the book (Chapters 4–8) revises some important tools of multivariable control theory and deals with relevant problems of MIMO process control. It aims at improving the closed-loop decoupling degree in the presence of either physical (actuator saturations), structural (decentralised controllers) or dynamic (non-minimum phase characteristics) constraints. There is also room to cope with bumpy transfers in MIMO processes.

The first MIMO problem to be treated is the dynamic decoupling preservation of multivariable processes in the presence of plant input constraints (Chapter 5). As already seen, input saturation changes the amplitude and the direction of the control signal that is necessary to achieve dynamic decoupling. Hence, in addition to the known problem of windup, the control directionality problem appears, bringing about the loss of the decoupling obtained for the ideal unconstrained case. Most of the existing methods to deal with this problem successfully avoid the change of control directionality by conditioning the whole reference vector. However, we here address the problem from a different perspective: to preserve the closed-loop decoupling as long as possible. Therefore, the presented approach does not affect those variables whose set points remain constant, avoiding in this way the generation of undesired transients in these channels.

As a second multivariable problem of major interest for industrial engineers, the reduction of cross-interactions in multiloop or decentralised control is considered in Chapter 6. The great majority of industrial process control loops for MIMO systems are still based on this architecture. However, in spite of their practical benefits, multiloop controllers are not able to suppress interactions. An illustrative example is given in the chapter to reveal the effects of crossed interactions on multiloop control. When the process coupling is significant, the pairing problem of choosing which available plant input is to be used to control each of the plant outputs must first be properly solved. Nonetheless, neither appropriate control structure selection nor controller tuning is sufficient to guarantee amplitude delimitation of the input–output coupling. The conditioning technique developed in Chapter 2 is shown in this chapter as a powerful tool to impose user-defined boundaries for the loop interactions in decentralised control systems, i.e. to robustly respect output signal bounds. All the chapter content is then applied to a benchmark quadruple tank process, whose dynamics simply represents the behaviour of several chemical and industrial processes.

A half-way strategy between diagonal decoupling and decentralised control is addressed in Chapter 7 for non-minimum phase processes, regarding that for these systems the former strategy spreads RHP zeros (thus imposing additional closed-loop performance constraints), while the latter does not generally give satisfactory results for relatively demanding closed-loop requirements. Hence, partial decoupling is considered and studied throughout the chapter. In a partially decoupled control system, non-zero off-diagonal elements in the closed-loop transfer matrix help to relax the bounds on sensitivity functions imposed by the RHP zeros, and also permit pushing the effects of these zeros to a particular output. However, the off-diagonal elements also give rise to interactions in a structured form, which strongly depend on the RHP zero directions. The conditioning algorithm is here propounded, analysed and designed to limit the remaining interactions in partial decoupled systems while avoiding undershoots in the (decoupled) variable of interest as much as possible. The same non-linear quadruple tank process of Chapter 6 is considered as a case study, but subjected to much more demanding control specifications.

Finally, Chapter 8 presents an algorithm for the reduction of the undesired effects caused by manual–automatic or controller switching in multivariable process control. Of course, the method is also applicable to SISO systems as a particular case. It simply uses a switching device and a first-order filter to avoid inconsistencies between the off line controller outputs and the plant inputs. As a consequence, jumps at the plant inputs are prevented (i.e. bumpless transfer is achieved) and undesired transients on controlled variables are significantly reduced. Some advantages of this method are its straightforward implementation, its robustness properties and that, unlike other bumpless proposals, it does not need a model of the plant.

Chapter 2
A practical method to deal with constraints

This chapter presents the main ideas of the book to deal with constraints in feedback control systems. The proposed method aims at preserving the simplicity of conventional anti-windup (AW) algorithms, but at the same time it is supported by a solid theoretical background and exhibits distinctive robustness features. Among them, the method does not require an exact model of the plant (it only needs knowledge of the model structure), it can be employed to address constraints in non-linear processes and it can transparently deal with a wide variety of non-linearities apart from amplitude saturation, such as rate limiter, asymmetry, dead zone, and time dependency. For the sake of simplicity, only single-input single-output (SISO) systems are considered in this chapter.

2.1 Introduction

In this chapter, we describe and analyse a unified and practically oriented method to address several constrained control problems. The methodology to be presented is the key topic behind the book's control approaches, since from the basic and simplest idea the algorithm is then extended to deal with different types of constraints in both SISO and multiple-input multiple-output (MIMO) systems.

The method recovers the idea of shaping the reference signal, which, as mentioned in Section 1.5, is one of the possible schemes of two-step algorithms for dealing with constraints. Here, we combine reference-conditioning ideas with sliding regime properties to improve robustness with respect to both the model of the constrained system and output disturbances. This greatly extends the applicability of the method as well. The implementation and design is simple enough for being applied to real industrial problems, whereas the corresponding analysis gives rigorous conditions for the proper operation of the method.

With the aim of keeping the presentation as simple as possible, we first describe conceptually how the method operates when it is applied to controllers or subsystems described by biproper transfer functions (see Section 2.2). After a simple example, some basic concepts on variable structure systems (VSS), necessary for the theoretical study of the algorithm, are presented in Section 2.4. The technique is then extended to deal with strictly proper controllers or subsystems in Section 2.5. From Section 2.6, the chapter generalises the method application to non-linear systems and discusses some implementation and robustness issues.

2.2   Preliminary definitions

First of all, we need to recall some simple but relevant definitions related to dynamical systems.

Definition 2.1 (Proper and improper systems): *A linear dynamical system represented by a transfer function $P(s)$ is proper if $\lim_{s\to+\infty}|P(s)|$ is finite. A proper system is in turn strictly proper provided $\lim_{s\to+\infty}|P(s)| = 0$ and biproper if $\lim_{s\to+\infty}|P(s)| > 0$. All systems that are not proper are improper.*

From the above definition, a system $P(s)$ is proper if the order of the numerator polynomial n_N is smaller than or equal to the order of the denominator polynomial n_D, otherwise it is improper. Note that improper transfer functions cannot be physically realised.

Definition 2.2 (Relative degree): *It is said that a proper system $P(s)$ has relative degree ρ when the difference between the system denominator order $n_D \equiv n$ and the numerator order n_N is $\rho = n_D - n_N$.*

In this way, the relative degree of a given system indicates the number of times that the system output must be differentiated for the system input to explicitly appear.

Definition 2.3 (Minimum phase): *A system $P(s)$ is non-minimum phase (NMP) if its transfer function contains zeros in the right-half plane (RHP) or time delays. Otherwise, the system is minimum phase (MP).*

2.3   Sliding mode reference conditioning

2.3.1   Basic idea for biproper systems

Figure 2.1 presents a block diagram of a simple scheme to deal with constraints in dynamical systems. Therein, one can distinguish a constrained dynamical system $S_c(s)$ to which an auxiliary loop was added to avoid breaking the limits $\underline{v}$ and $\bar{v}$ on its constrained variable v.

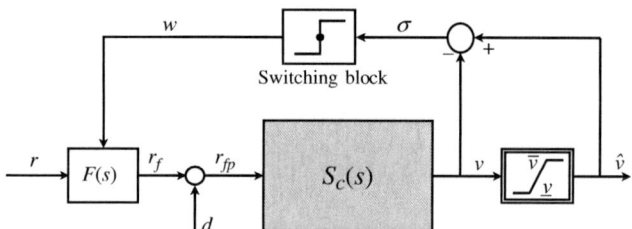

Figure 2.1   Block diagram of a conditioning loop to deal with constraints in biproper systems

It is important to remark that the system $S_c(s)$ generically represents a constrained subsystem of the whole control system, and therefore the variable v may correspond to any system variable subjected to constraints. For instance, $S_c(s)$ may stand for a biproper controller facing input saturation (i.e. $v \equiv u$, u is the controller output), or a strictly proper feedback control system subjected to output signal bounds ($v \equiv y$). In the latter case, the states of the constrained subsystem will be necessary for the conditioning loop, as we will see further in Section 2.5 (Figure 2.9).

Now, we focus on the scheme of Figure 2.1 and assume $S_c(s)$ is a linear, biproper, stable and MP system (e.g. a PI controller). Therefore, its dynamic behaviour can be represented by

$$S_c(s) : \begin{cases} \dot{x}_s = A_s x_s + b_s r_{fp} \\ v = c_s^{\mathrm{T}} x_s + d_s r_{fp} \end{cases} \tag{2.1}$$

with $x_s \in X_s \subset \mathbb{R}^n$, $A_s \in \mathbb{R}^{n \times n}$, b_s, $c_s \in \mathbb{R}^n$ and $d_s \in \mathbb{R}$, $d_s \neq 0$. Its perturbed input is

$$r_{fp} = r_f + d \tag{2.2}$$

where d is a disturbance signal.

A signal $\hat{v}$, which always accomplishes the specified constraints on v, can be generated by means of an artificial *limiting element* (depicted with double box in Figure 2.1) as follows:

$$\hat{v} = \begin{cases} \bar{v} & \text{if} \quad v > \bar{v} \\ v & \text{if} \quad \underline{v} \leq v \leq \bar{v} \\ \underline{v} & \text{if} \quad v < \underline{v} \end{cases} \tag{2.3}$$

where the constant values $\bar{v}$ and $\underline{v}$ correspond to the upper and lower limits allowed for variable v. Observe that, although (2.3) represents amplitude saturation, any other typical non-linearity could have been considered.

The following commutation law is then implemented in the *switching block* of the auxiliary loop to fulfil the bounds $\underline{v}$ and $\bar{v}$:

$$w = \begin{cases} w^- & \text{if } \sigma < 0 \\ w^+ & \text{if } \sigma > 0 \\ 0 & \text{if } \sigma = 0 \end{cases} \tag{2.4}$$

with w^-, $w^+ \in \mathbb{R}$, $w^- \neq w^+$ and the switching function σ defined as

$$\sigma = \hat{v} - v \tag{2.5}$$

The resulting discontinuous signal w is employed to shape the system input r_f through a first-order low-pass filter represented by the block

$$F(s) : \begin{cases} \dot{x}_f = \lambda_f x_f + w + r \\ r_f = -\lambda_f x_f \end{cases} \tag{2.6}$$

where r is the original system reference and $\lambda_f \in \mathbb{R}$ is the filter eigenvalue. Naturally, this eigenvalue must be chosen for the filter bandwidth to be much faster than the constrained dynamical system in such a way that the system response is not deteriorated when the output v is within its allowed range.

Observe that (2.3) and (2.5) determine a linear region

$$\mathcal{R} = \{x_s \in X_s : \sigma = \hat{v} - v = 0\} \tag{2.7}$$

and two limiting surfaces

$$\begin{aligned}\bar{\mathcal{S}} &= \{x_s \in X_s : \bar{\sigma} = \bar{v} - v = 0\} \\ \underline{\mathcal{S}} &= \{x_s \in X_s : \underline{\sigma} = \underline{v} - v = 0\}\end{aligned} \tag{2.8}$$

When the system operates within its allowed region $\mathcal{R}(\underline{v} \le v \le \bar{v})$, the signal w is zero and no reference correction is performed, i.e. the conditioning loop is inactive. However, when v tries to exceed its upper bound $\bar{\mathcal{S}}$, which makes $\bar{\sigma} < 0$, the signal w changes to w^-. Similarly, if v intends to fall below its lower limit $\underline{\mathcal{S}}$, making $\underline{\sigma} > 0$, w switches to w^+. This situation is depicted in Figure 2.2 for amplitude limits on v.

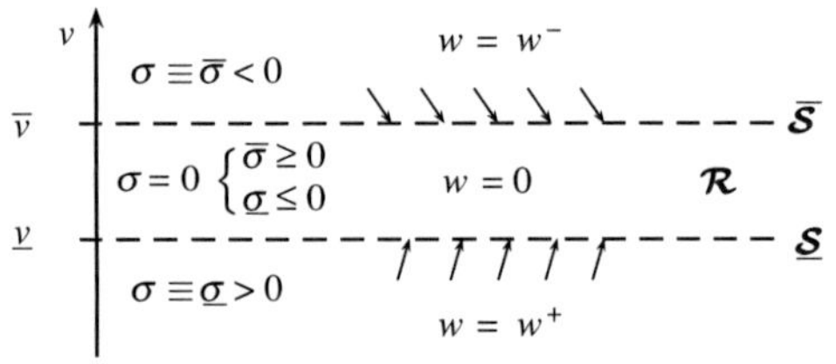

Figure 2.2   Graphical interpretation of switching law (2.4) and (2.5)

As can be noticed from (2.1), (2.2) and (2.6), the switching signal w directly affects the first time derivative of the constrained signal v and the switching function σ. Thus, for a sufficiently large discontinuous signal w, the switching logic ((2.4) and (2.5)) ensures that

$$\begin{cases} \dot{\bar{\sigma}} > 0 \ \text{if} \ \bar{\sigma} < 0 \\ \dot{\underline{\sigma}} < 0 \ \text{if} \ \underline{\sigma} > 0 \end{cases} \tag{2.9}$$

In this manner, when v intends to surpass one of its limits, the commutation law ((2.4) and (2.5)) will make w non-zero to enforce v re-entering its allowed region, where w equals zero again. Whenever the variable v continues trying to cross the system output bounds, the signal w will be switching between 0 and w^- (or w^+) at high frequency because of (2.9), and the system will evolve so that $\bar{\sigma} = 0$ (or $\underline{\sigma} = 0$). Consequently, the conditioned signal r_f will be continuously adjusted in such a way that the output v never exceeds the predefined limits.

Remark 2.1: *The method does only employ a switching device and a first-order filter, and even then it has many attractive properties that will be shown and illustrated throughout the book. Among them, it does not depend on the system model (only on its relative degree), it is robust against uncertainty and external* *disturbances, it is confined to the low-power side of the system and thus easy to implement, etc.*

Remark 2.2: *Because of the switching nature of the algorithm, well-established VSS theory can be used to analyse most of the method features, which is the subject of Section 2.4. It will also allow generalising the method to deal with strictly proper linear systems (Section 2.5) and non-linear systems as well (from Section 2.6).*

Remark 2.3: *The high-frequency switching of w, which results from the commutation law (2.4) and (2.5), can be interpreted as a transient sliding regime on the surfaces $\overline{S}$ or $\underline{S}$ during r_f adjustment. For this reason, from now on we will refer to this methodology as sliding mode reference conditioning (SMRC).*

2.3.2   Illustrative example

Consider again the Example 1.1 given in Chapter 1. To avoid controller windup, we apply the SMRC scheme just described, which for the case of biproper controllers is as simple as shown in Figure 2.3.

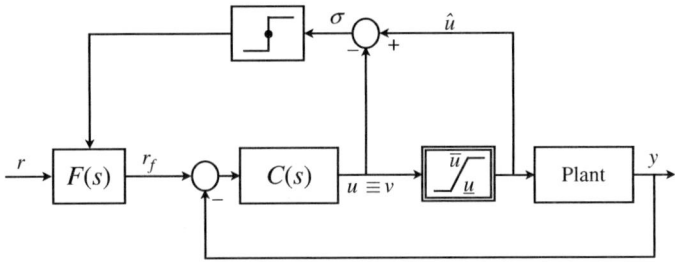

Figure 2.3   SMRC as anti-windup algorithm for biproper controllers

Note that in this case the constrained variable v coincides with the main control action u. Also, the subsystem $S_c(s)$ is here given by the closed-loop transfer function between $r_f + d$ and u, with the disturbance d given by the perturbed component of y. For greater details on this output decomposition, see Section 4.2.2 in Chapter 4. Observe that $S_c(s)$ is biproper because the PI controller in (1.3) is also biproper.

The eigenvalue of the filter (2.6) is taken as $\lambda_f = -1$, which makes the filter much faster than the closed loop. The switching signal w commutes according to (2.4) and (2.5). Note that the switching function $\sigma = \hat{u} - u$ has unitary relative degree with respect to w, since the error e (whose first derivative depends on w) is implicit in u through the proportional term of the PI controller.

The response of the closed-loop system with the SMRC technique as AW method is presented in Figure 2.4 for the same amplitude limits as the ones considered in Example 1.1. On the one hand, the evolution of the control action u and the plant input $\hat{u}$ reveals that $u \equiv \hat{u}$ all the time, i.e. the actuator never saturates. On the other hand, the response and settling time of the controlled variable y greatly improve with respect to the non-compensated case (see Figure 1.3). The dotted lines depict once more the ideal unconstrained responses for comparative purposes.

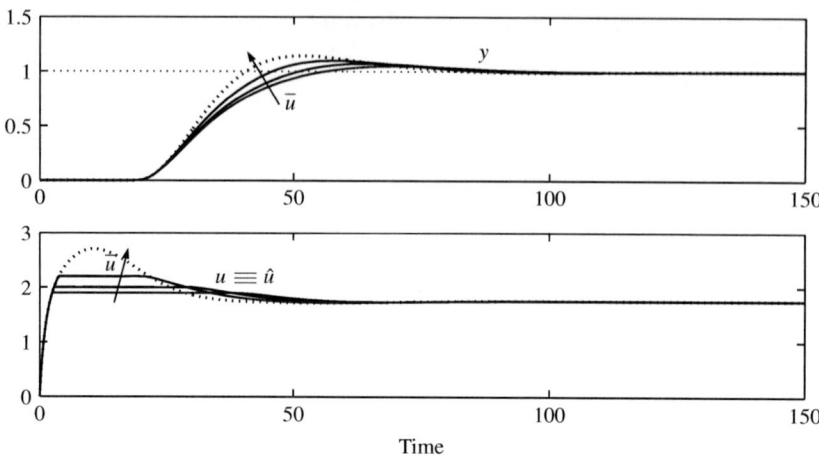

Figure 2.4   Conditioned system: output y and control action u for increasing values of ū

Figure 2.5 shows the conditioned reference signals r_f and the discontinuous signals w corresponding to Figure 2.4. As could be expected, as the bound on the control action $\bar{u}$ increases, the conditioned reference tends to the original reference command.

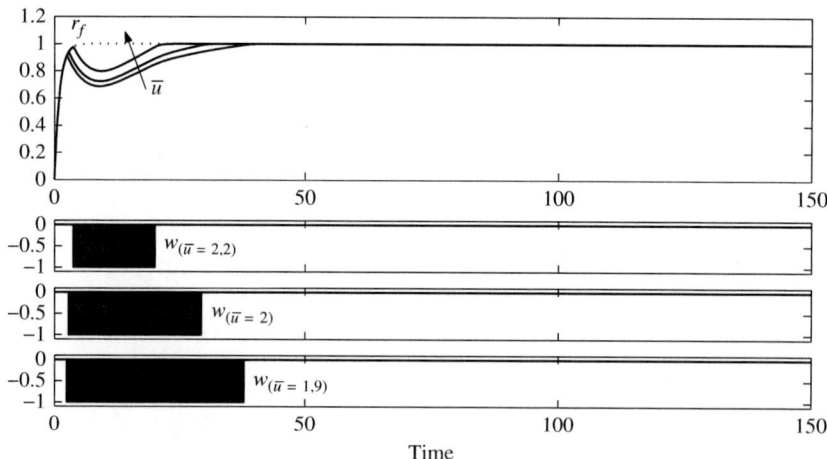

Figure 2.5   Filtered reference and discontinuous action corresponding to Figure 2.4

Finally, in Figure 2.6 the time evolution of u and y together with the system trajectories on the plane (u, x_i) is plotted for the case $\bar{u} = 2$, with x_i being the

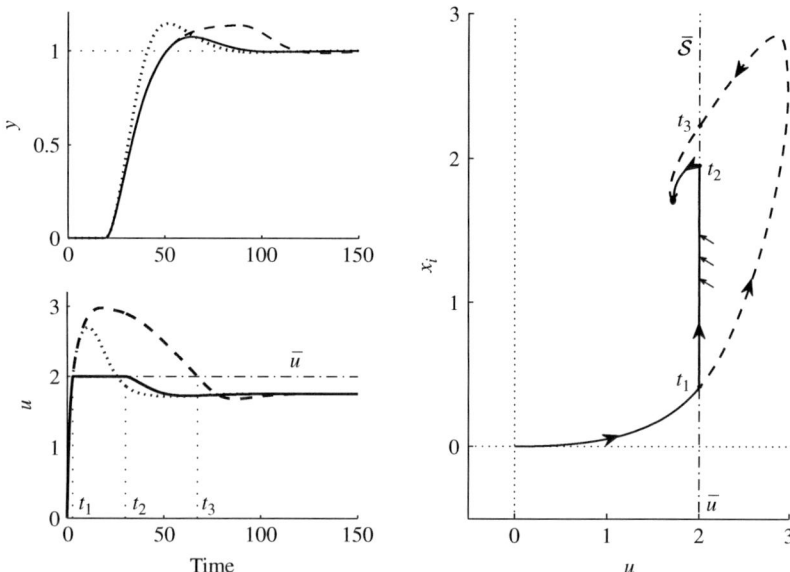

Figure 2.6 *Time evolution of u and y and system trajectories on the plane (u, x_i)*
for the case $\bar{u} = 2$ without AW compensation (dashed) and with SMRC
compensation (solid). Dotted line: ideal unconstrained case

integral state of the PI controller. In the absence of windup compensation, the
control signal u enters the forbidden saturated region at instant t_1, re-entering
linear operation only after instant t_3. During the interval $t_1 - t_3$, an important
growth of the integral state x_i, typical of the windup effect, can be observed.
This gives rise to a long settling time in the controlled variable y. The trajectory
of the compensated system coincides with the previous one until instant t_1. From
then on, the SMRC loop produces consecutive abrupt changes in the trajectory
direction by means of the discontinuous signal w, which forces the control action
to re-enter the linear region. This situation is repeated at high frequency, and a
so-called sliding regime is established on the surface $\bar{\sigma} = 0$. In t_2, the system
trajectory stops pointing outside the linear region. Thus, the conditioning loop
becomes inactive and the system evolves with its own linear dynamics to the
equilibrium point.

2.4 Biproper SMRC: features and analysis

As was already mentioned, VSS-related concepts can be used to formalise and
extend the methodology described in the previous section. To this end, some basics
about VSS and sliding regimes are presented first. Then, they are employed for the
analysis of the SMRC algorithm when dealing with biproper constrained subsystems and to gain insight into the method properties.

2.4.1   VSS essentials

A VSS is a dynamical system composed of various continuous subsystems with a switching logic. In a VSS, the system structure is intentionally changed with, among others, the following objectives:

- To improve closed-loop performance.
- To solve control problems that continuous control is not able to work out.
- To attain robustness against model uncertainties and external disturbances.

The discontinuous control action resulting from the switching logic of a VSS system achieves a particular operation when switching occurs at a very high frequency, constraining the system state to a surface on the state space. This kind of operation is called sliding mode and has many attractive properties. Among them, it is robust to parameter uncertainties and external disturbances, the closed-loop system is an order-reduced one and its dynamics depends on the designer-chosen sliding surface [114,135]. Because of its interesting features, a large number of works presenting practical applications of SM have been reported during the past decades in the main journals on control systems.

From the beginning of VSS studies [29,133], great attention has been paid to the development of a theoretical framework to deal with these kinds of switching systems, which are typically described by differential equations with discontinuous right-hand side. Nowadays, VSS are considered to be theoretically well founded. Let us now review some fundamental concepts of VSS and sliding regimes. The interested reader is referred to books [26,136] that are completely devoted to SM control and its applications.

2.4.1.1   Sliding mode description

Consider a continuous linear system given by

$$P(s) : \begin{cases} \dot{x} = Ax + bu \\ y = c^{\mathrm{T}}x \end{cases} \tag{2.10}$$

where $x \in X \subset \mathbb{R}^n$ is the state vector, $u \in \mathbb{R}$ is the (discontinuous) control action, $A \in \mathbb{R}^{n \times n}$, b and $c \in \mathbb{R}^n$.

If a switching function $\sigma(x)$ is defined as a smooth function $\sigma : X \to \mathbb{R}$, whose gradient $\nabla \sigma$ is non-zero in X, the set

$$\mathcal{S} = \{x \in X : \sigma(x) = 0\} \tag{2.11}$$

defines a regular manifold in X of dimension $n - 1$, which is called sliding manifold or surface.

Such a surface can be reached by properly defining a variable structure control law, which makes the control action u to take one of two possible values, depending on the sign of the switching function $\sigma(x)$:

$$u = \begin{cases} u^+ & \text{if } \sigma(x) > 0 \\ u^- & \text{if } \sigma(x) < 0 \end{cases} \quad u^+ \neq u^- \tag{2.12}$$

Whenever the switching law (2.12) enforces the system to reach the surface $\mathcal{S}$ and to remain locally around it, it is said that there exists a sliding regime on the surface $\mathcal{S}$. To this end, the vector fields controlled by the two continuous subsystems, $(Ax + bu^+)$ and $(Ax + bu^-)$, must point locally towards the manifold $\mathcal{S}$. This situation is graphically represented by Figure 2.7.

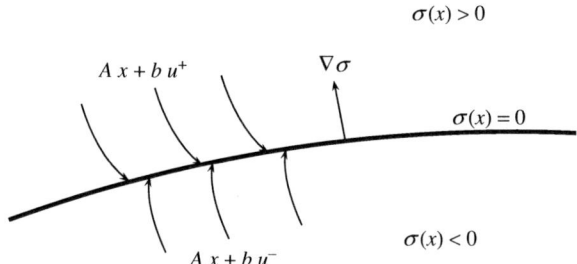

Figure 2.7   Sliding regime establishment on the surface $\sigma(x) = 0$

2.4.1.2   Necessary conditions for SM establishment

We now look for necessary conditions so that the situation depicted by Figure 2.7 holds. Mathematically, it can be described by (note the similarities with (2.9))

$$\begin{cases} \dot{\sigma}(x) < 0 \ \text{ if } \ \sigma(x) > 0 \\ \dot{\sigma}(x) > 0 \ \text{ if } \ \sigma(x) < 0 \end{cases} \tag{2.13}$$

Equation (2.13) implies that the rate of change of the scalar function $\sigma(x)$ always opposes the sign of $\sigma(x)$, which guarantees crossing $\mathcal{S}$ from both sides of the surface. The same condition around $\mathcal{S}$ can be written in compact form as

$$\lim_{\sigma(x) \to 0} \sigma(x) \cdot \dot{\sigma}(x) < 0 \tag{2.14}$$

From (2.10) and (2.12), we can also express (2.13) as

$$\dot{\sigma}(x) = \left(\frac{\partial \sigma}{\partial x}\right)^{\mathrm{T}} \dot{x} = \begin{cases} \left(\dfrac{\partial \sigma}{\partial x}\right)^{\mathrm{T}} Ax + \left(\dfrac{\partial \sigma}{\partial x}\right)^{\mathrm{T}} bu^+ < 0 \ \text{ if } \ \sigma > 0 \\[2ex] \left(\dfrac{\partial \sigma}{\partial x}\right)^{\mathrm{T}} Ax + \left(\dfrac{\partial \sigma}{\partial x}\right)^{\mathrm{T}} bu^- > 0 \ \text{ if } \ \sigma < 0 \end{cases} \tag{2.15}$$

Then, it is obvious that for SM to establish on $\sigma(x) = 0$, the inequality

$$\left(\frac{\partial \sigma}{\partial x}\right)^{\mathrm{T}} b \neq 0 \tag{2.16}$$

must hold locally on $\mathcal{S}$. Condition (2.16) is a necessary condition for SM establishment, and it is generally known as *transversality condition* [26].

As a particular case, consider the following switching law

$$\sigma(x) = k_r r - k^T x \qquad (2.17)$$

where the constant k_r is chosen so that the steady-state value of the output equals the set point r, and the feedback gains k^T determine the linear dynamics during SM (see Figure 2.8).

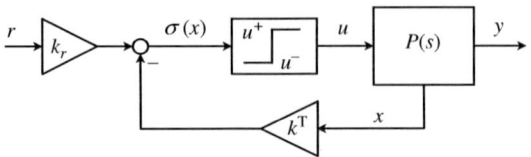

Figure 2.8   *Conventional SM control scheme*

Clearly, the transversality condition for the switching function (2.17) is given by $k^T b \neq 0$. Moreover, the transfer function from input u to σ results

$$H_{SM}(s) = -k^T(sI - A)^{-1}b \qquad (2.18)$$

which can be written as

$$H_{SM}(s) = (-k^T)bs^{-1} + (-k^T)Abs^{-2} + \cdots + (-k^T)A^m bs^{-(m+1)} + \cdots \qquad (2.19)$$

by applying the Taylor's series decomposition. Therefore, the transversality condition imposes that the first term of the Taylor's series decomposition must be non-zero ($k^T b$ is the first Markov parameter of the corresponding linear system). This means that the transfer function from the discontinuous control action to the sliding function σ must have unitary relative degree.

2.4.1.3   Continuous equivalent control

From a theoretical point of view, during SM operation the system switches at infinite frequency. It is then discontinuous at every time instant, and its state equation does not admit an analytical solution in the classical sense. Alternatively, a simple and useful way of computing the system dynamics during SM consists of finding an equivalent continuous control signal.

With this aim, the *ideal sliding mode* is defined as an ideal operating mode for which the manifold S is a system invariant. Thus, once the system trajectory reaches the surface, it slides exactly along S. The invariance condition of manifold S is given by

$$\sigma(x) = 0 \qquad (2.20)$$

$$\dot{\sigma}(x) = \left(\frac{\partial \sigma}{\partial x}\right)^T (Ax + bu_{eq}) = 0 \qquad (2.21)$$

Equation (2.21) states that the trajectory remains on the surface, whereas $u_{eq}(x)$ represents a continuous *equivalent control* action for which S is a local invariant manifold of system (2.10). The function $u_{eq}(x)$ can be easily derived from (2.21), yielding

$$u_{eq} = -\left[\left(\frac{\partial \sigma}{\partial x}\right)^{\mathrm{T}} b\right]^{-1} \left(\frac{\partial \sigma}{\partial x}\right)^{\mathrm{T}} Ax \qquad (2.22)$$

As can be seen, the transversality condition (2.16) is a necessary and sufficient condition for the equivalent control to be well defined. Note also that u_{eq} makes sense only on the surface $\sigma(x) = 0$.

For the particular case of sliding function (2.17), the equivalent control is obviously given by

$$u_{eq} = -(k^{\mathrm{T}} b)^{-1} k^{\mathrm{T}} Ax \qquad (2.23)$$

2.4.1.4 Necessary and sufficient condition for SM

We have already seen that the transversality condition is necessary to accomplish condition (2.13), and therefore for sliding regime to exist. Now, from the u_{eq} definition we can derive a necessary and sufficient condition to guarantee SM establishment. In effect, by subtracting (2.21) from the first line of (2.15) we have

$$\left(\frac{\partial \sigma}{\partial x}\right)^{\mathrm{T}} b(u^+ - u_{eq}) < 0 \qquad (2.24)$$

while

$$\left(\frac{\partial \sigma}{\partial x}\right)^{\mathrm{T}} b(u^- - u_{eq}) > 0 \qquad (2.25)$$

results from taking away (2.21) from the second line of (2.15). Then, from (2.24) and (2.25) and assuming without loss of generality that $u^+ > u^-$, the following necessary and sufficient condition yields:

$$\begin{cases} \left(\frac{\partial \sigma}{\partial x}\right)^{\mathrm{T}} b < 0 \\ u^- < u_{eq} < u^+ \end{cases} \qquad (2.26)$$

Note that if $\left(\frac{\partial \sigma}{\partial x}\right)^{\mathrm{T}} b > 0$ or $u^+ < u^-$ condition (2.26) is still valid, but for a new switching function $\sigma^*(x) \overset{\Delta}{=} -\sigma(x)$.

2.4.1.5 SM reduced dynamics

The resultant reduced dynamics during ideal SM can be obtained by substituting in the state equation of system (2.10) the input u with the equivalent control action (2.22), which gives

$$\dot{x} = Ax + b\left\{-\left[\left(\frac{\partial \sigma}{\partial x}\right)^{\mathrm{T}}b\right]^{-1}\left(\frac{\partial \sigma}{\partial x}\right)^{\mathrm{T}}Ax\right\} \qquad (2.27)$$

Reordering,

$$\dot{x} = \left\{I - b\left[\left(\frac{\partial \sigma}{\partial x}\right)^{\mathrm{T}}b\right]^{-1}\left(\frac{\partial \sigma}{\partial x}\right)^{\mathrm{T}}\right\}Ax \qquad (2.28)$$

which for the linear switching function (2.17) and its corresponding equivalent control (2.23) reduces to

$$\dot{x} = \underbrace{[I - b(k^{\mathrm{T}}b)^{-1}k^{\mathrm{T}}]A}_{A_{SM}}x \qquad (2.29)$$

The state equations (2.28) and (2.29) redundantly describe the system dynamics during SM. In fact, one of their rows is linearly dependent on the remaining $n-1$ equations because the system state x satisfies the algebraic constraint $\sigma(x) = 0$. Then, the matrix A_{SM} has an eigenvalue at the origin, which has to be attributed to this redundant description and does not imply sliding regime instability.

2.4.2   SMRC operation analysis

Given the earlier review of VSS fundamentals, we are now interested in analysing the SMRC method described in Section 2.3 from a VSS point of view. This will allow us to show some important features of the algorithm.

2.4.2.1   SM operation in SMRC

By comparing the SMRC method of Section 2.3 with conventional SM control, a first significant difference appears: whereas in conventional VSS control schemes SM is the desired mode of operation and is sought from both sides of the sliding surface (Figure 2.7), the SMRC approach defines two limiting surfaces, which delimit the desired linear operation region and which are not sought (Figure 2.2). Instead, they are reached only if the own trajectories of the system lead the variable v to come to one of its limits. In fact, since the switching law (2.4) makes the discontinuous signal zero at one side of each surface, SMRC could be thought of as producing a 'one-side SM', in which the system slides on the surface only if it continues trying to exceed the bounds by itself. In this manner, SM is only a transient mode of operation to avoid the system surpassing its constraints (assuming sustainable references, see subsection 2.4.2.6).

2.4.2.2   Necessary condition for SMRC

As we have seen, for SM to be established the transversality condition (2.16) must hold, i.e. the first time derivative of the switching function must explicitly depend on the discontinuous control action. Differentiating the sliding function (2.5) of the

SMRC approach, it is straightforward from (2.1) and (2.6) to derive that this necessary condition is satisfied provided

$$d_s \lambda_f \neq 0 \tag{2.30}$$

Thus, although the simple algorithm of Section 2.3.1 is useful to illustrate the main idea of the method operation, it is only valid for systems having a direct path from the input to the constrained variable v (i.e. biproper systems with $d_s \neq 0$). In the case of strictly proper constrained systems $S_c(s)$, additional system states should be included in the switching function so that the transversality condition holds. That is, other sliding function rather than the trivial one of (2.5) should be defined whose first derivative explicitly depends on the discontinuous action w. This case is addressed in Section 2.5.

2.4.2.3 Equivalent control

The continuous equivalent signal to the high-frequency switching signal produced when the system reaches one of its limits can be obtained from the invariance conditions, generically given by (2.20) and (2.21). For the SMRC algorithm, this condition corresponds to $\tilde{\sigma} = 0$ and $\dot{\tilde{\sigma}} = 0$, where the tilde is used to denote either the upper or the lower bar ($\tilde{\star} = \overline{\star}$ or $\tilde{\star} = \underline{\star}$). Using the system and filter representations (2.1) and (2.6), it can be easily shown that

$$\tilde{w}_{eq} = r_f - r - \frac{1}{\lambda_f}\dot{d} + \frac{1}{d_s \lambda_f}(\dot{v} - c_s^T \dot{x}_s)\Big|_{\tilde{\sigma}=0} \tag{2.31}$$

It is evident that the continuous equivalent control is only well defined provided $d_s \neq 0$, which confirms the arguments given in the previous paragraph.

2.4.2.4 Necessary and sufficient condition for SMRC

From the general condition (2.26) and the particular switching logic of the SMRC algorithm (2.4), the following conditions can be derived for SM establishment on each of the limiting surfaces:

$$0 < \underline{w}_{eq} < w^+ \tag{2.32}$$

$$w^- < \overline{w}_{eq} < 0 \tag{2.33}$$

where $w^- < 0 < w^+$ was assumed and both $\underline{w}_{eq}$ and $\overline{w}_{eq}$ are given by (2.31).

It is important to remark that the inequalities with respect to zero in (2.32) and (2.33) are verified whenever the dynamical system attempts, by its own, to cross the bounds $\overline{v}$ or $\underline{v}$. When this occurs, the discontinuous signal amplitudes w^- or w^+ should be large enough to satisfy the remaining inequalities, so that SM is guaranteed and the system constraints are fulfilled. According to (2.31), this can always be achieved for appropriate bounds on $\dot{d}$, r, $\dot{v}$ and $\dot{x}_s$.

It is also worthy of mention in this regard that the selection of w^+ and w^- can be made in a conservative manner because the SM loop is restricted to the low-power side of the system. Indeed, they could be implemented either via analog electronics or as numerical values in a microprocessor algorithm.

2.4.2.5   Reduced and robust SMRC dynamics

The state equations of the whole conditioned system can be grouped as

$$
\begin{bmatrix} \dot{x}_s \\ \dot{r}_{fp} \end{bmatrix} = \begin{bmatrix} A_s & b_s \\ 0 & -\lambda_f \end{bmatrix} \begin{bmatrix} x_s \\ r_{fp} \end{bmatrix} + \begin{bmatrix} 0 \\ \lambda_f r + \lambda_f d + \dot{d} \end{bmatrix} + \begin{bmatrix} 0 \\ \lambda_f w \end{bmatrix} \tag{2.34}
$$

However, during the transient sliding regimes the surface $\sigma(x) = 0$ establishes an algebraic relation between these states. In effect, since v coincides with $\tilde{v}$ on the limiting surfaces, from the second line of (2.1) the following equality holds:

$$
r_{fp} = d_s^{-1}(\tilde{v} - c_s^{\mathrm{T}} x_s) \tag{2.35}
$$

where d_s^{-1} is finite because the system was assumed biproper. If the previous expression of r_{fp} is replaced in the first line of (2.34), the *reduced* dynamics of the system during sliding regime operation is found[1]:

$$
\begin{aligned}
\dot{x}_s &= Q_s x_s + b_s d_s^{-1} \tilde{v} \\
Q_s &= (A_s - b_s d_s^{-1} c_s^{\mathrm{T}})
\end{aligned} \tag{2.36}
$$

Observe that, provided (2.32) and (2.33) hold, during SMRC $v \equiv \tilde{v}$ independently of the reference r and the disturbance d at the input of the constrained subsystem $S_c(s)$. Such robustness features are based on the so-called *matching condition* of sliding regimes, which states that the dynamics during SM is insensitive to any parameter being collinear with the discontinuous action. Although this condition is shown in Section 2.7, it is evident from (2.34) that the input r and the disturbance d satisfy the matching condition.

As $v \equiv \tilde{v}$, (2.36) determines the *hidden* dynamics during SMRC. It is easy to show that the eigenvalues of Q_s are the zeros of the system (2.1). Therefore, the internal dynamics of the conditioning loop will be globally stable provided $S_c(s)$ is MP, as was originally assumed. If this were not the case, the conditioning algorithm could only be locally applied (see also Remark 5.1).

2.4.2.6   Condition for transient SMRC operation

A sufficient condition for the reestablishment of unconstrained operation (inactive SMRC loop) can be derived from (2.31). Assume that SM is established on surface $\bar{\sigma} = \tilde{v} - v = 0$, which means that the continuous equivalent control satisfies $w^- < \overline{w}_{eq} < 0$. The system recovers the nominal closed-loop dynamics provided

---

[1] Note that the resulting SMRC dynamics could have been obtained also by replacing w in (2.34) with the w_{eq} expression of (2.31). Indeed, this leads the second row in (2.34) to be equal to the derivative of (2.35).

$\overline{w}_{eq}$ reaches 0. Thus, we need to study the asymptotic behaviour of $\overline{w}_{eq}$ to find under which conditions it will be greater than 0 in finite time. Considering constant amplitude limits for v and disturbances with zero steady-state derivative, if the system continued undergoing SM on $\overline{\sigma} = 0$, $\overline{w}_{eq}$ would tend towards

$$\lim_{t \to \infty} \overline{w}_{eq} = \lim_{t \to \infty} (r_f - r) \tag{2.37}$$

In addition, since $v = \overline{v}$ during SM

$$\lim_{t \to \infty} r_f = (d_s - c_s^{\mathrm{T}} A_s^{-1} b_s)^{-1} \overline{v} = S_c(0)^{-1} \overline{v} \tag{2.38}$$

where $S_c(0)$ is the DC-gain of system (2.1). Therefore, for step references $r < S_c(0)^{-1} \overline{v}$ the asymptotic value $\lim_{t \to \infty} \overline{w}_{eq}$ will be greater than 0, which guarantees that SMRC becomes inactive in finite time.

Reasoning in the same fashion for $\underline{\sigma} = 0$ yields the following sufficient condition for the reestablishment of unconstrained operation:

$$|r| < |S_c(0)^{-1} \tilde{v}| \tag{2.39}$$

2.4.3   Implementation issues

Another distinctive feature of SMRC is that it does not present the two main drawbacks of VSS practical applications, namely the *chattering* effect and the *reaching mode*.

2.4.3.1   Chattering

In conventional SM applications, performance is usually deteriorated by undesired oscillations of finite frequency and amplitude. This effect is known as chattering and is produced when the system state does not slide along the sliding surface, but it oscillates around $\sigma = 0$ (inside a given 'band').

There are basically two main sources of chattering: the presence of unmodelled parasitic dynamics and the own limitations of the switching devices, such as actuator delays or computing times. Much research has been devoted to the study and the attenuation of chattering-related problems [30,119,136,134], including the currently active line on high-order sliding modes [6,32,76,77].

An interesting feature of the SMRC algorithm is that SM is confined to the low-power side of the system, where fast electronic devices can be used to implement the discontinuous action. In effect, the algorithm could be implemented in a few lines of microprocessor routine or even via analog electronics using a dual operational amplifier chip. Thus, differing from conventional SM control, in SMRC schemes the switching frequency has not to be bounded to protect plant actuators or mechanical devices. In addition, in SMRC one can always include r_{fp} (or a signal containing it) in the switching function, ensuring in this manner the unitary relative degree despite the unmodelled dynamics. Therefore, both causes of chattering are significantly reduced, making this problem negligible in most SMRC practical implementations.

2.4.3.2   Reaching mode

The reaching mode is the operation phase of a VSS during which the state trajectory evolves from an initial condition towards a point on the sliding manifold where the SM existence conditions hold. The reaching condition is given by (2.13) and must be guaranteed by means of the control law (2.12).

Although there exist in the literature several methods for the operation during the reaching phase [64], a common characteristic of all of them is that during this operation mode the switching signal does not commute but it just takes a given (open-loop) value depending on the employed method. This clearly differs from the desired SM operation, where the discontinuous signal switches at high frequency to enforce the system to evolve on the prescribed sliding surface. In effect, the attractive robustness properties of SM strategies are only present during SM operation, but not during the reaching phase. Consequently, the reaching mode may degrade the global performance of a VSS [84].

However, in the SMRC methodology discussed in this chapter, there is no reaching mode because of its operation principle. Since the desired mode of operation falls within system limits, no control action is applied to drive the system state towards the sliding surfaces. Instead, they are reached only if the own trajectories of the system are likely to violate the unavoidable bounds, leading to transient SM establishment on the limiting surface to avoid it.

2.5   Strictly proper SMRC

Our objective now is to generalise the SMRC algorithm of Section 2.3 to deal with constrained systems $S_c(s)$ of arbitrary relative degree ρ. Since the particular case of biproper constrained systems ($\rho = 0$) was already addressed, in this section we will only consider strictly proper systems. Note that for this kind of systems $d_s = 0$, and so the trivial switching function (2.5) does not satisfy the corresponding transversality condition (2.30). Therefore, as already warned, other system states should be considered in the switching functions to guarantee their unitary relative degree with respect to the discontinuous action w. This is schematically represented in Figure 2.9.

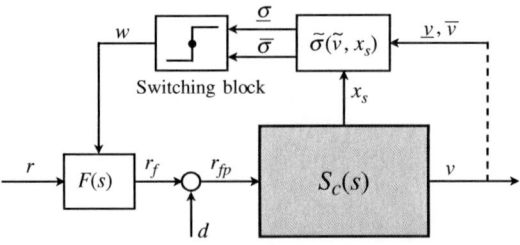

Figure 2.9   Block diagram of an SMRC loop for strictly proper constrained systems

2.5.1 Normal form

From a practical viewpoint, any state-space model can be employed to include the system states in the SMRC loop. However, the description provided by the normal canonical form [65] allows a conceptually simple demonstration of the algorithm properties. Thus, we here remind how to compute such a model for a linear strictly proper system.

Assume the constrained system (2.1) has relative degree $1 \leq \rho \leq n$ between the output v and the input r_{fp}. The ρ successive derivatives of the output are given by

$$
\begin{cases}
v & = c_s^T x_s \\
\dot{v} & = c_s^T \dot{x}_s = c_s^T A_s x_s + c_s^T b_s r_{fp} \\
\ddot{v} & = c_s^T A_s \dot{x}_s = c_s^T A_s^2 x_s + c_s^T A_s b_s r_{fp} \\
& \vdots \\
v^{(\rho-1)} & = c_s^T A_s^{(\rho-1)} x_s + c_s^T A_s^{(\rho-2)} b_s r_{fp} \\
v^{(\rho)} & = c_s^T A_s^{(\rho)} x_s + c_s^T A_s^{(\rho-1)} b_s r_{fp}
\end{cases}
\tag{2.40}
$$

where the first $(\rho - 1)$ Markov parameters $c_s{}^T b_s = c_s{}^T A_s b_s = \cdots = c_s{}^T A_s{}^{(\rho-2)} b_s = 0$ and the ρth one $c_s{}^T A_s{}^{(\rho-1)} b_s \neq 0$ because of the relative degree assumption. Hence, only the last equation of (2.40) explicitly depends on the perturbed conditioned input r_{fp}.

Then, by applying a linear transformation $v = [\xi_s^T \ \eta_s^T]^T = T x_s$ to system (2.1), with

$$
T = \begin{bmatrix}
c_s^T \\
c_s^T A_s \\
c_s^T A_s^2 \\
\vdots \\
c_s^T A_s^{(\rho-1)} \\
\tau_{\rho+1} \\
\vdots \\
\tau_n
\end{bmatrix}
\tag{2.41}
$$

$\tau_{p+1}, \ldots, \tau_n$ being arbitrarily chosen row vectors such that T is non-singular, its dynamics can be conveniently expressed in the normal canonical form:

$$S_c(s) : \begin{cases} v_1 &= c_s^T x_s = v \\ \dot{v}_1 &= c_s^T A_s x_s = v_2 \\ \dot{v}_2 &= c_s^T A_s^2 x_s = v_3 \\ &\vdots \\ \dot{v}_{\rho-1} &= c_s^T A_s^{(\rho-1)} x_s = v_\rho \\ \dot{v}_\rho &= c_s^T A_s^{(\rho)} x_s \big|_{x_s = T^{-1} v} + c_s^T A_s^{(\rho-1)} b_s r_{fp} \\ \dot{\eta}_s &= P_s \xi_s + Q_s \eta_s. \end{cases} \tag{2.42}$$

where $\xi_s = [v_1 v_2 \ldots v_\rho]^T$ comprises the system output v and its first $(\rho - 1)$ derivatives and η_s are $(n - \rho)$ linearly independent states.

2.5.2   Method reformulation

From (2.42), the switching law (2.4) and the sliding function (2.5) are reformulated as follows for constrained systems with $\rho \geq 1$:

$$w = \begin{cases} w^- & \text{if } \bar{\sigma} < 0 \\ w^+ & \text{if } \underline{\sigma} > 0 \\ 0 & \text{otherwise} \end{cases} \tag{2.43}$$

with

$$\begin{cases} \bar{\sigma} = \bar{v} - v - k^T \dot{\xi}_s = \bar{v} - v - \sum_{\alpha=1}^{\rho} k_\alpha v^{(\alpha)} \\ \underline{\sigma} = \underline{v} - v - k^T \dot{\xi}_s = \underline{v} - v - \sum_{\alpha=1}^{\rho} k_\alpha v^{(\alpha)} \end{cases} \tag{2.44}$$

and where $k = col(k_\alpha)$ is a vector of constant gains with $k_\rho \neq 0$. The transversality condition for SM establishment ($\bar{\sigma}$ having relative degree one with respect to w) can be easily verified by recalling that $v^{(\rho)}$ is proportional to r_{fp} according to (2.40), and that $\dot{r}_{fp}$ depends on w because of (2.2) and (2.6). In fact, the whole state-space description of the conditioning loop is given by (2.42) and

$$\dot{r}_{fp} = -\lambda_f (r_{fp} - d - r - w) + \dot{d} \tag{2.45}$$

The switching policy represented by (2.43) and (2.44) also states that when there is no risk of leaving the allowed region, $w = 0$ and the conditioning loop is inactive. However, SMRC now starts when the output v gets close to one of its limits, without necessarily reaching it. This is because (2.44) considers not only the amplitude of the constrained variable v but also its subsequent derivatives. For example, if $\rho = 1$ the speed with which v tends to its bound is also taken into account to start the reference conditioning.

2.5.2.1 Reduced and robust SMRC dynamics

The reduced dynamics during SMRC operation can be easily derived from (2.42) by making the switching functions in (2.44) equal zero. This leads to

$$
\begin{cases}
v_1 & = \ v \\
\dot{v}_1 & = \ v_2 \\
\dot{v}_2 & = \ v_3 \\
\cdots & = \ \cdots \\
\dot{v}_{\rho-1} & = \ v_\rho \\
\dot{v}_\rho & = \ \left(\tilde{v} - v - \sum_{\alpha=1}^{\rho-1} k_\alpha v_{\alpha+1} \right) \Big/ k_\rho \\
\dot{\eta}_s & = \ P_s \xi_s + Q_s \eta_s
\end{cases}
\tag{2.46}
$$

where the state equation (2.45) has been removed because of redundancy.

Hence, during SMRC the system output v evolves according to the first ρ lines of (2.46), with the constants k_α being the coefficients of the characteristic poly-nomial from the bound $\tilde{v}$ to the variable v. Thus, assuming a canonical model (based on a chain of integrators) is available, the dynamics with which v will tend towards its maximum values will only depend on the switching function design. This robustness property of the SMRC scheme will also be analysed in Section 2.7.

It is important to remark that to avoid breaking the bounds on variable v, the coefficients k_α must always be chosen for the output dynamics to be overdamped. Moreover, by properly selecting the values of k_α, a desired approaching rate can be set to avoid hard-hitting the bounds.

Observe also that in the normal canonical form (2.46), the zeros of the system (2.1) are the eigenvalues of Q_s, and so they determine the hidden dynamics during SMRC. In this manner, as already claimed in Section 2.4 for biproper systems, global stability of SMRC algorithm is only guaranteed for MP strictly proper sys-tems $S_c(s)$. However, we will see in further chapters that this restriction does not prevent SMRC from being applied to non-minimum phase MIMO systems, since multivariable transmission zeros may be in the RHP while all the individual transfer functions are of MP. In the case of non-minimum phase SISO systems, SMRC should be applied locally on an invariant region, or an alternative MP output should be sought so that the system constraints are indirectly fulfilled when this output is bounded.

Remark 2.4: *Actually, $v^{(\alpha)}$ in (2.44) are not to be computed by differentiating v, which may be unadvisable in practice due to the noise components in the output* signals. Instead, they can be generated from the system input r_{fp} and states x_s by means of the linear transformation (2.41). Although this transformation depends on the constrained system parameters, this is not critical for proper operation of SMRC. Indeed, the model parameters only affect the transient dynamics during SMRC, which should not be exactly known but it only has to be overdamped. In addition, the limits $\tilde{v}$ can always be set conservatively to account for uncertainty.

2.5.3   Illustrative examples

2.5.3.1   Output or state limiter for a closed-loop system

We consider a closed-loop system with a PI controller like the one in Figure 2.10. The whole system can be represented in compact form by the following equations:

$$\dot{x} = \begin{bmatrix} 0 & 0.01 & 0 \\ 0 & -1 & 1 \\ 0 & 0 & 0 \end{bmatrix} x + \begin{bmatrix} 0 \\ 50 \\ 0.001 \end{bmatrix} (r_f - y) \tag{2.47}$$

$$y = x_1 \tag{2.48}$$

The first two lines of (2.47) describe the dynamics of the process under control, whereas the third one is the integral state of the PI controller. It is supposed here that, for safe operation of the process, the variable x_2 must be upper bounded to $\bar{x}_2 = 20$.

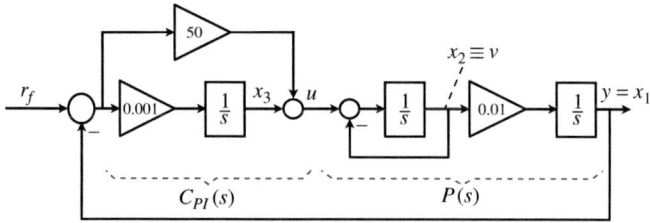

Figure 2.10   *SMRC for delimiting internal state*

To fit within the general framework previously developed, we should consider $v = x_2$ and $\bar{v} = \bar{x}_2$. The output-constrained system $S_c(s)$ of (2.1) is given by the two last lines of (2.47), followed by a remaining system described by the first differential equation. Note that the system $S_c(s)$ is then strictly proper.

To meet the constraint on $v = x_2$, an SMRC loop is employed. Unfortunately, the obvious switching law $\sigma(x) = x_2 - \bar{x}_2$ cannot be used because the transversality condition $d_s \lambda_f \neq 0$ does not hold (see (2.30)). So, a sliding regime cannot be established on $\sigma(x) = 0$ and x_2 might cross the surface and evolve towards dangerous operating regions. To overcome this obstacle, a new switching law defined by

$$\sigma(x) = \bar{x}_2 - x_2 - k_1 \dot{x}_2 \tag{2.49}$$

is proposed. Note that the reference is implicit in $\dot{x}_2 = -x_2 + 50r_f - 50y + x_3$, and therefore the sliding function (2.49) has unitary relative degree with respect to the discontinuous action, provided the added filter is of first order. Thus, such a reference filter is included in the auxiliary conditioning loop.

Figure 2.11 plots the simulation results with (solid line) and without (dashed line) reference compensation. To ensure safe operation, the SMRC starts even before x_2 reaches its upper bound. During the sliding regime, x_2 converges exponentially towards its limit value with time constant k_1. Figure 2.11(a) shows the smooth convergence of x_2 to its bound at $\bar{x}_2 = 20$ with the fast sliding dynamics. Figure 2.11(c) displays the associated system trajectory in the phase plane $(x_2, \dot{x}_2)$. The SMRC protection acts along the line $t_1 - t_2$. Obviously, the slope of the sliding surface, as well as the sliding dynamics, is given by the sliding gain k_1. Once the system trajectory points towards the safety region from both sides of the sliding surface, the system naturally evolves within this region without reference conditioning towards its equilibrium point. Finally, Figure 2.11(b) shows the graceful degradation of the controlled variable x_1 caused by the SMRC.

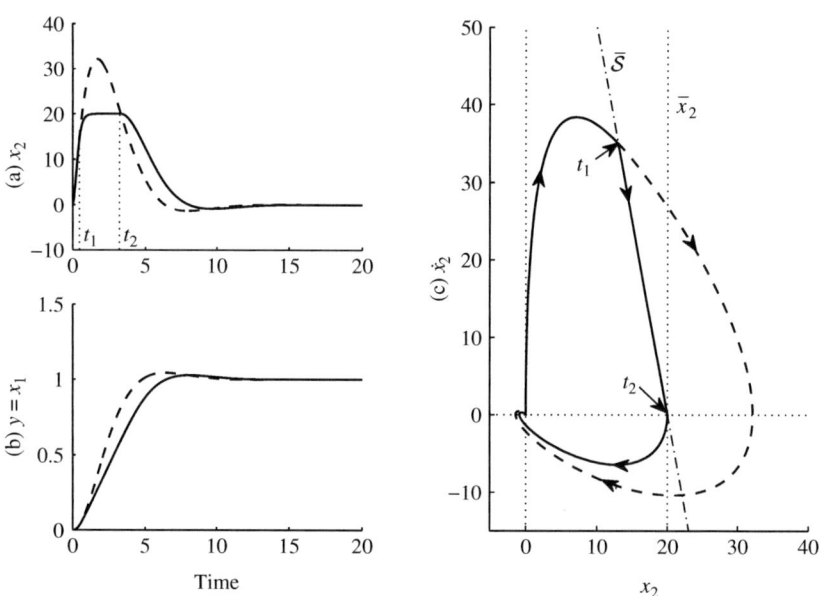

Figure 2.11 *Closed-loop response with (solid) and without (dashed) SM conditioning: (a) variable x_2, (b) variable x_1, and (c) phase trajectories in $(x_2, \dot{x}_2)$*

2.5.3.2 Prevention of plant windup

Rigorous methods to remove plant windup can be found in References 13, 60 and 129. Although they give strict proof of stability, Reference 129 requires the complete model of the plant to be added as a dynamic element to the controller, whereas References 13 and 60 make use of non-linear stability tools like the circle criterion, which generally lead to conservative designs. Here, making use of SMRC properties, we adopt a simpler but less rigorous approach, which may result interesting in practical applications.

We bring back Example 1.2 given in Chapter 1. Recall that plant windup was caused by amplitude of the system states that cannot be reduced fast enough. Figure 1.4 revealed how the slow reaction of state x_3 because of input saturation gives rise to underdamped (or even oscillatory) output responses. Then, our objective here is to delimit the amplitude of this first state to avoid its undesired consequences. Seen as constrained output, x_3 has relative degree of one with respect to the plant input u (see (1.6)). We therefore propose the switching functions:

$$\tilde{\sigma}(x) = \tilde{x}_3 - x_3 - k_1 \dot{x}_3 = \tilde{x}_3 - x_3 + k_1(x_1 + 2x_2 + 3x_3 - u) \tag{2.50}$$

with $\tilde{x}_3 = \pm 0.1$ and $k_1 = 0.01$, which results in an SMRC time constant of 0.01 [time units]. The same precompensating filter (1.5) added to avoid steady-state error is employed to smooth out the discontinuous signal w generated by (2.43).

In this way, when the state x_3 is about to reach its predefine bound, the SMRC shapes the reference signal via the discontinuous action w in order to limit it. The corresponding results are shown in Figure 2.12 (solid lines), together with the original curves of Example 1.2 (dotted lines). Evidently, SMRC greatly improves the closed-loop responses, avoiding the undesired effect of plant windup.

2.5.3.3   Constrained unstable system

Undesired effects of controller and plant windup are much worse in unstable systems. Indeed, it is well known that open-loop unstable systems cannot be globally stabilised in presence of plant input constraints [126].[2] This is because the output variable may reach values from which the constrained control action is not able to bring it back. Thus, there is always a range of the output variable that has not to be exceeded in order to avoid the system to become unstable. Naturally, this restricts the reference amplitude and/or the initial conditions (to the so-called maximal output admissible sets [44]). Among other authors, Hippe [60] claims that the system output should not exceed $y_{max} \leq \bar{u} \cdot |P(0)|$, with $\bar{u}$ the input saturation limit and $P(0)$ the DC-gain of the unstable plant.

As a simple unstable example, consider the plant

$$P(s) = \frac{0.5}{s - 5} \tag{2.51}$$

---

[2] Observe indeed that this kind of systems gives rise to NMP (closed-loop) transfer functions $S_c(s)$ between the reference and the plant input, which is consistent with the analysis of Sections 2.4.2 and 2.5.2.

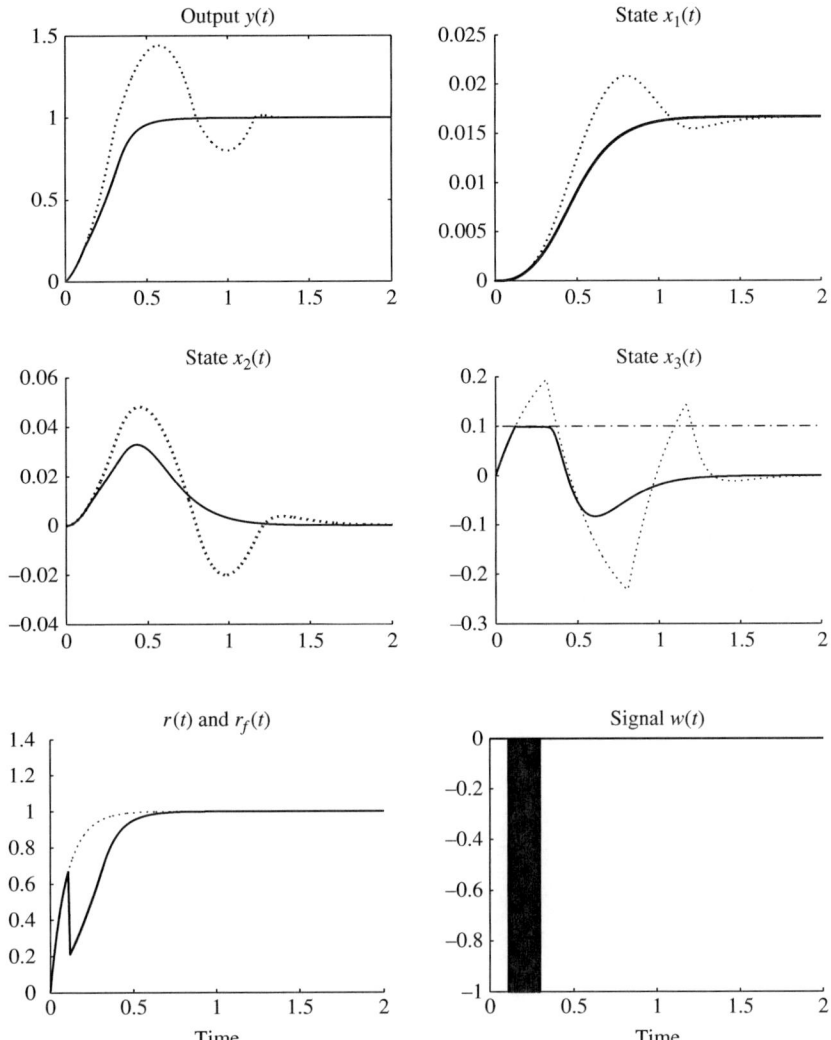

Figure 2.12   Plant windup prevention via internal state bounding with SMRC algorithm

with its input range limited to $\hat{u} \leq \bar{u} = 15$. The responses of this plant when it is controlled by the PI controller

$$C_{PI}(s) = \frac{100(s + 30)}{s} \tag{2.52}$$

without AW compensation are shown in Figure 2.13(a). As can be appreciated, the system effectively becomes unstable when y goes beyond 1.5. Because of the poor

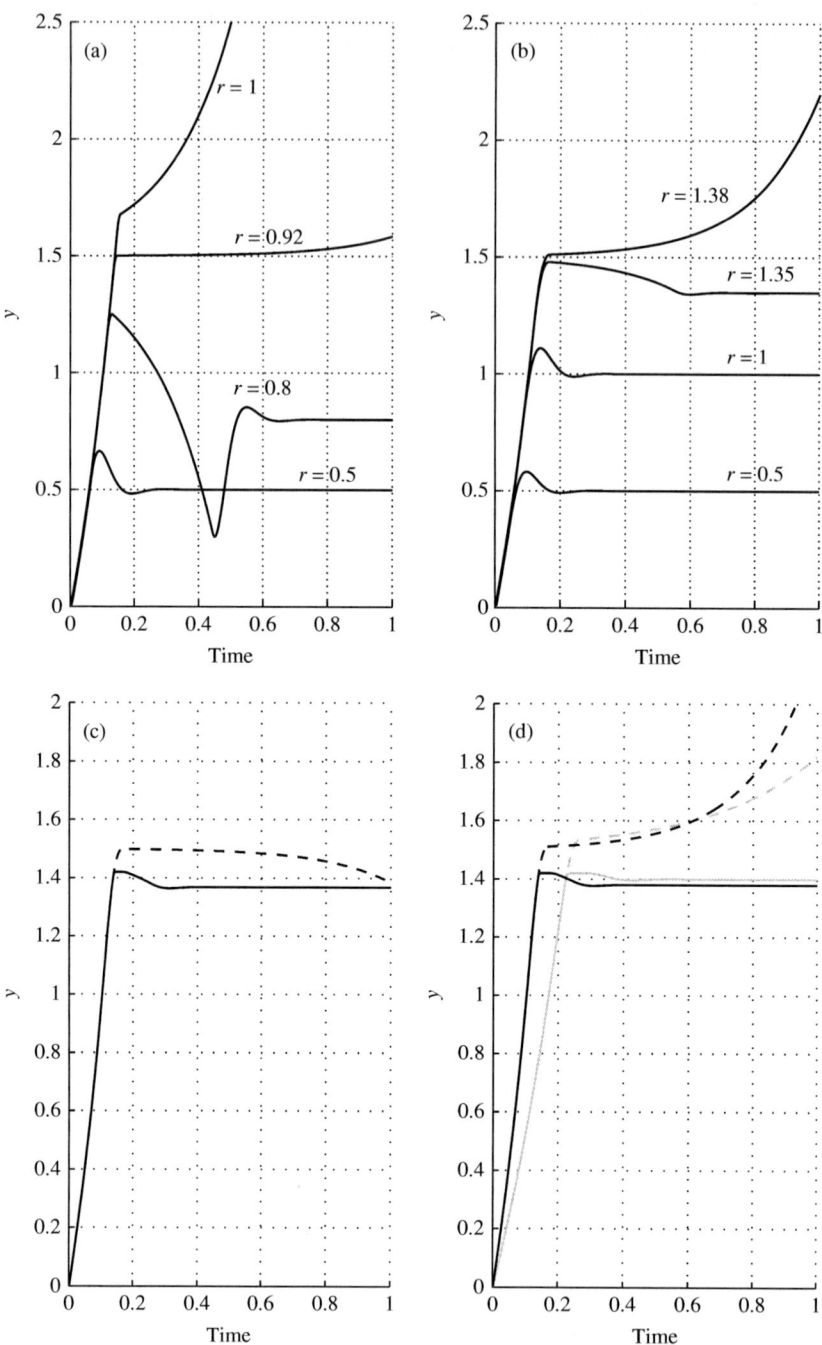

*Figure 2.13   Constrained unstable system responses: (a) uncompensated system,
(b) with AW compensation, (c) and (d) AW compensation (dashed)
vs. AW and output delimiting (solid)*

transient responses caused by controller windup, this occurs from reference steps just greater than 0.9.

Controller AW methods solve the inconsistencies between the controller states and the plant input, improving the transient responses and enlarging in this way the range of reference values that keep the system inside the allowed region. However, if $y > y_{max}$ the system still becomes unstable. This is verified in Figure 2.13(b) (note the greater reference values). Although an SMRC scheme was employed here as controller AW method, any other AW algorithm could have been used and the same conclusions would have been drawn.

Two methods are proposed in Reference 60 for overcoming this problem, where it is studied in greater detail. The first one merely consists of designing the loop sufficiently slow so as to guarantee that always $|y(t)| \leq r < y_{max}$. The second one is based on a feedback control for disturbance rejection and system stabilisation, plus a feedforward part with a non-linear model-based reference generator. This feedforward part makes use of ideas presented earlier in Reference 13.

We here evaluate SMRC to cope with constrained unstable systems. The idea consists of delimiting the system output ($y = v$) to a given value lower than y_{max}. As the closed-loop transfer function (which is in this case 'our' constrained system $S_c(s)$ has relative degree of one, the proposed switching function σ includes the first derivative of the output, with constant $k_1 = 0.002$. The discontinuous signal is smoothed out by a first-order filter faster than the closed loop.

The achieved results are presented in Figure 2.13(c) and (d) (solid lines) and compared with the responses obtained when only AW compensation is added (dashed lines). As can be observed at the bottom left box, a reference $r = 1.37$ leads the system without SMRC output delimiting just on the border of the allowed region. In fact, the plot (d) reveals that for a slightly larger reference step, the AW-compensated system becomes unstable. On the contrary, the SMRC output bounded system does not only present better closed-loop response for references $r \leq 1.37$ but also remains stable for greater set points, enlarging the set of admissible references.

It is important to highlight the SMRC robustness properties (recall Remark 2.4). On the one hand, the output bound can be conservatively set to account for uncertainty on the plant DC-gain (in the example we chose $\bar{y} = 1.42$ for a better visualisation). On the other hand, if $\dot{y}$ in σ is obtained from the canonical model of the plant, the SMRC dynamics is completely insensitive to the uncertainty. Assuming a canonical model is not accessible, $\dot{y}$ can be computed from the plant input as $\dot{y} = (u + 10y)/0.5$. Although it is now dependent on the model parameters, this only affects the SMRC time constant with which the output bound is reached, which is not critical at all for SMRC operation and should not be exactly known. This is verified by the grey curves in Figure 2.13(d), which show the responses ($r = 1.4$) obtained when the real system is forced to be $\tilde{P}(s) = 0.3/(s - 3)$ and the function σ of SMRC is computed from the nominal model (2.51). Actually, the difference in the fast SMRC time constant (0.002 [time units] for the nominal case) is not even appreciable despite the great change in the system dynamics.

2.6   SMRC and non-linear systems

The geometric interpretation of the SMRC method not only helps understand its operation principles and robustness features but also simplifies its generalisation to deal with constrained non-linear systems. With this aim, we first present a geometric approach of sliding regimes over non-linear systems, and from this analysis we then explore the connections between SMRC and geometric invariance concepts.

2.6.1   *Geometrical interpretation of SM*

Let us consider the non-linear system

$$\begin{cases} \dot{x} = f(x) + g(x)u \\ y = h(x) \end{cases} \tag{2.53}$$

where $x \in X \subset \mathbb{R}^n$ and $u \in \mathbb{R}$ play the same role as in system (2.10), i.e. u is given by (2.12), $f : \mathbb{R}^n \to \mathbb{R}^n$ and $g : \mathbb{R}^n \to \mathbb{R}^n$ are vector fields in C^n (infinitely differentiable) and $h(x) : \mathbb{R}^n \to \mathbb{R}$ a scalar field also in C^n, all of them defined in X, with $g(x) \neq 0, \forall x \in X$.

For our geometric analysis purposes, we introduce the *directional derivative* or *Lie derivative*

$$L_f h(x) : \mathbb{R}^n \to \mathbb{R}$$

which denotes the derivative of a scalar field $h(x) : \mathbb{R}^n \to \mathbb{R}$ in the direction of a vector field $f(x) : \mathbb{R}^n \to \mathbb{R}^n$

$$L_f h(x) = \frac{\partial h}{\partial x} f(x)$$

Note that since $L_f h(x)$ is a scalar function, the Lie derivative can be recursively applied

$$L_f^k h(x) = \frac{\partial}{\partial x}(L_f^{k-1} h(x)) f(x)$$

This allows a compact notation of directional derivatives, both for a single vector field or several ones. For example, considering two vector fields $f(x)$ and $g(x)$

$$L_g L_f h(x) = \frac{\partial}{\partial x}(L_f h(x)) g(x)$$

Now, we can differentiate the switching function $\sigma(x)$, which defines the manifold (2.11) in the direction of the trajectories of the non-linear system (2.53). As the Lie derivative is also a linear operator, we have

$$\dot{\sigma}(x) = L_{f+gu}\sigma = L_f \sigma + L_g \sigma u \tag{2.54}$$

The control law (2.12) leads, in turn, to SM operation on the surface $\sigma(x) = 0$. So, from the invariance condition defined in Section 2.4, ($\sigma(x) = 0$, $\dot{\sigma}(x) = 0$) the equivalent control results

$$u_{eq}(x) = -\frac{L_f \sigma}{L_g \sigma}\bigg|_{\sigma=0} \tag{2.55}$$

Replacing u in (2.53) by u_{eq} yields

$$\dot{x} = f(x) + g(x)u_{eq} = f(x) - \frac{L_f \sigma}{L_g \sigma}g(x) \tag{2.56}$$

The ideal SM dynamics can therefore be obtained by substituting the Lie derivatives and reordering (2.56):

$$\dot{x} = \left[I - g(x)\left(\frac{\partial \sigma}{\partial x}g(x)\right)^{-1}\frac{\partial \sigma}{\partial x}\right]f(x) = F(x)f(x) \tag{2.57}$$

We have seen that the continuous action u_{eq} leads the system to evolve on the sliding surface $\mathcal{S}$. According to this, the dynamics $F(x)f(x) = f(x) + g(x)u_{eq}$, which results from applying u_{eq} is tangent to $\mathcal{S}$, and thus normal to the gradient of $\sigma(x)$ ($\nabla\sigma = \partial\sigma/\partial x$). That is,

$$F(x)f(x) \in \ker(\nabla\sigma) \tag{2.58}$$

Consequently, we can see $F(x)$ as an operator that projects the vector $f(x)$ onto the plane tangent to the surface $\mathcal{S}$. To find in which direction $F(x)$ performs this projection, let $z(x)$ be a collinear vector with $g(x)$ ($z \in span(g)$) of arbitrary amplitude:

$$z(x) = g(x)\mu(x), \text{ with } \mu(x) : \mathbb{R}^n \to \mathbb{R} \tag{2.59}$$

By applying $F(x)$ to $z(x)$ we get

$$F(x)z(x) = \left[I - g(x)\left(\frac{\partial \sigma}{\partial x}g(x)\right)^{-1}\frac{\partial \sigma}{\partial x}\right]g(x)\mu(x) = 0 \tag{2.60}$$

The cancellation of (2.60) means that $F(x)$ projects any vector in the direction of the control field $g(x)$, i.e. along $span(g)$.

Figure 2.14 depicts this geometric approach. Observe that $F(x)f(x)$ is the projection of $f(x)$ onto $\mathcal{S}$ in the direction of $g(x)$, and thus, the control u_{eq} is such that $F(x)f(x)$ is tangent to $\mathcal{S}$.

Remark 2.5: *Figure 2.14 also reveals the geometric implications of the transversality condition, which from (2.53) and (2.54) is given by*

$$L_g\sigma = \frac{\partial \sigma}{\partial x}g(x) \neq 0 \tag{2.61}$$

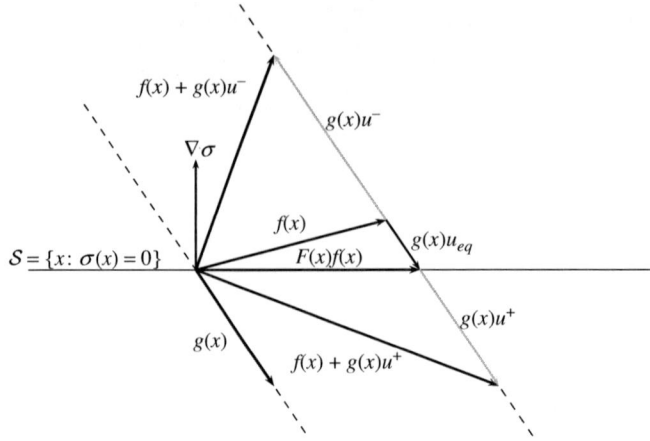

Figure 2.14   Geometrical interpretation of sliding mode dynamics

It clearly states that the field $g(x)$ cannot be tangent to the sliding manifold $\mathcal{S}$, i.e. $g(x) \notin ker(\nabla \sigma)$.

2.6.2   Geometric invariance via SMRC

Geometric invariance ideas have been used in the literature to derive stability conditions for hardly constrained non-linear control systems [85,149]. We next explore the relationship of the SMRC approach with geometric invariance control and their corresponding invariance conditions.

Consider an alternative constrained output of system (2.53)

$$\begin{cases} \dot{x}_s = f(x_s) + g(x_s)u \\ v = h_v(x_s, u) \end{cases} \tag{2.62}$$

with the variable v denoting the real valued system variable that has to be bounded so as to fulfil user-specified system constraints. To specify the bounds on v, the following set can be defined:

$$\Phi(x_s) = \{x_s | \phi(v) \leq 0\} \tag{2.63}$$

where standard geometric invariance notation has been preserved.[3] Naturally, the trivial case would be

$$\phi(v) = v - \tilde{v} \tag{2.64}$$

The goal then is to find a control input u such that the region Φ becomes invariant (i.e. trajectories originating in Φ remain in Φ for all times t). To ensure the invariance

---

[3] Note that ϕ plays the same role as $\tilde{\sigma}$ in the previous sections.

of Φ, the control input u must guarantee that the right-hand side of the first equation in (2.62) points to the interior of Φ at all points on the border surface:

$$\partial\Phi = \{x_s | x_s \in \Phi \wedge \phi(v) = 0\} \tag{2.65}$$

This situation is geometrically depicted in Figure 2.15. Mathematically, it can be expressed by the scalar product as

$$\|\nabla\phi\| \|f + gu\| \cos\theta = \frac{\partial\phi}{\partial x_s}\dot{x}_s = \dot{\Phi}(x_s, u) \leq 0, \forall x_s \in \partial\Phi \tag{2.66}$$

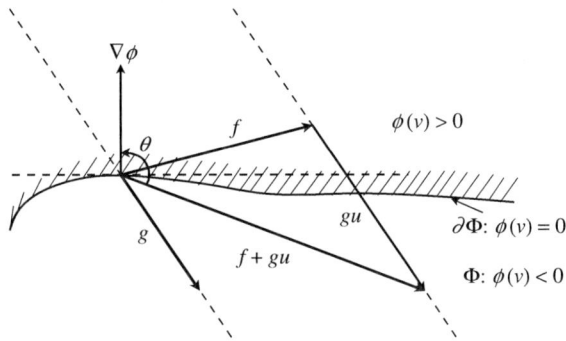

Figure 2.15 Invariance condition $\cos(\theta) < 0$

Clearly, this requires that at least the minimum of $\dot{\Phi}(x_{s,b}, u)$ with respect to u for each constant $x_{s,b} \in \partial\Phi$ is non-positive. If the maximum value of the resulting set of minima (i.e. the *infimum*) is non-positive, then invariance control is feasible. Hence, (2.66) can be written in standard form as

$$\inf_u \dot{\Phi}(x_s, u) \leq 0, \quad \text{with } x_s \in \partial\Phi \tag{2.67}$$

which is known as *implicit invariance condition* [1,85]. Solving (2.67) for u gives rise to the *explicit invariance condition* for system (2.62)

$$u = \begin{cases} \leq u^\phi & : x_s \in \partial\Phi \wedge L_g\phi > 0 \\ \geq u^\phi & : x_s \in \partial\Phi \wedge L_g\phi < 0 \\ \nexists & : x_s \in \partial\Phi \wedge L_g\phi = 0 \\ \text{free} & : x_s \in \Phi \setminus \partial\Phi \end{cases} \tag{2.68}$$

with $u^\phi = -L_f\phi/L_g\phi$ and where $L_f\phi > 0$ was assumed without loss of generality.

The control u in the interior of Φ can be freely assigned. Essentially, two possibilities arise depending on the problem to be dealt with. On the one hand, it could be designed $u = 0$ so that the system evolves autonomously throughout the interior of Φ. Then, the control action becomes active only when the critical constraint is reached, i.e. when the state trajectory reaches the border $\partial\Phi$ trying to leave the set Φ. For those cases in which pushing the system to its limits is aimed, a more aggressive option would be choosing u in such a way that the state trajectories are enforced to reach the border $\partial\Phi$, i.e. the process is driven towards its critical constraint, and then to remain on it.

According to both (2.4) and (2.43), the SMRC approach adopts the former strategy. However, there are no limitations to adopt other switching criteria, as can be seen for instance in the robotic application in Chapter 3. Particularly, the switching policy ((2.4) and (2.5)) can be seen as making the set Φ invariant by means of the following law for each limiting function ($\underline{\sigma}$ or $\bar{\sigma}$):

$$u = \begin{cases} u^+ & \text{if } \phi(v) > 0 \\ 0 & \text{if } \phi(v) \leq 0 \end{cases} \quad \text{with } u^+ \neq 0 \tag{2.69}$$

This aims to satisfy

$$\dot{\phi}(x_s, u) = \begin{cases} L_f\phi + L_g\phi u^+ < 0 & \text{if } \phi(v) > 0 \\ L_f\phi > 0 & \text{if } \phi(v) < 0 \end{cases} \tag{2.70}$$

locally around $\partial\Phi$, so that a sliding regime is established on surface $\partial\Phi$. As already mentioned, the second inequality in (2.70) is locally satisfied whenever the system tries by itself to leave the set Φ. Thus, the switching law (2.69) does not seek for SM, but it establishes if the process is at the border of the allowed region and about to leave it. Additionally, the first inequality in (2.70) implies that for a sliding regime to be established on $\phi(v) = 0$ the transversality condition

$$L_g\phi = \frac{\partial\phi}{\partial x_s}g(x_s) \neq 0 \tag{2.71}$$

must hold locally on surface $\partial\phi$. If this is not the case, an auxiliary subset $\Phi^* \subset \Phi$ satisfying this condition should be properly defined based on the switching functions of Section 2.5. This is shown in the next subsection.

It is interesting to compare the equivalent control during sliding mode (2.55) with the necessary control for geometric invariance u^ϕ, considering the equivalence between ϕ and σ. According to (2.68), u_{eq} coincides with the control required to keep the system just on the border surface $\partial\Phi$. Consequently, the SMRC generated by switching law (2.69) produces the minimal change in the system trajectory in order to achieve the set $\Phi(x_s, u)$ to be robustly invariant. Moreover, the necessary condition for SM (2.71) guarantees that the invariant control exists in (2.68).

2.6.3   SMRC in strictly proper non-linear systems

Assuming now the constrained system S_c is strictly proper

$$S_c : \begin{cases} \dot{x}_s = f(x_s) + g(x_s) r_{fp} \\ v = h_v(x_s) \end{cases} \qquad (2.72)$$

with relative degree $1 < \rho \leq n$, we must follow an analogous procedure to the one of Section 2.5. A non-linear transformation $v = \begin{bmatrix} \xi_s^T & \eta_s^T \end{bmatrix}^T = T(x_s)$, smooth and invertible, is defined as

$$v = T(x_s) = \begin{bmatrix} h(x_s) \\ L_f h_v(x_s) \\ \vdots \\ L_f^{(\rho-1)} h_v(x_s) \\ \tau_{\rho+1}(x_s) \\ \vdots \\ \tau_n(x_s) \end{bmatrix} \qquad (2.73)$$

The functions $\tau_k(x_s)$ with $\rho + 1 \leq k \leq n$ can be arbitrarily chosen, with the condition that $T(x_s)$ remains being a diffeomorphism (non-singular). Hence, the normal form results

$$S_c : \begin{cases} v_1 & = v \\ \dot{v}_1 & = v_2 \\ \dot{v}_2 & = v_3 \\ & \vdots \\ \dot{v}_{\rho-1} & = v_\rho \\ \dot{v}_\rho & = b(\xi_s, \eta_s) + a(\xi_s, \eta_s) r_{fp} \\ \dot{\eta}_s & = q(\xi_s, \eta_s) + p(\xi_s, \eta_s) r_{fp} \end{cases} \qquad (2.74)$$

where $\xi_s = [v_1 \ v_2 \ldots v_\rho]^T$ and $\eta_s = [v_{\rho+1} \ v_{\rho+2} \ldots v_n]^T$. The dependence of variables η_s on r_{fp} can be removed by properly choosing the components τ_k with $\rho + 1 \leq k \leq n$ [65]. We assume in the following that this was carried out.

Thus, the output dynamics (v and its derivatives) can be set by the design of the switching function as in (2.44). We then propose for this non-linear case

$$\tilde{\sigma}(\xi) = \tilde{v} - v - k^T \dot{\xi}_s = \tilde{v} - v - \sum_{\alpha=1}^{\rho} k_\alpha v^{(\alpha)} \qquad (2.75)$$

where $k = col(k_\alpha)$ is again a vector of constant gains with $k_\rho \neq 0$. This switching function satisfies the transversality condition for SM establishment since $v^{(\rho)}$ is proportional to r_{fp} (see (2.74)), and $\dot{r}_{fp}$ is given by

$$\dot{r}_{fp} = -\lambda_f(r_{fp} - d - r - w) + \dot{d} \tag{2.76}$$

because of the first-order filter of SMRC schemes.

During SMRC $\tilde{\sigma}(\xi) = 0$, and therefore the reduced SMRC dynamics is

$$\dot{\xi}_s \begin{cases} \dot{v}_1 & = & v_2 \\ \dot{v}_2 & = & v_3 \\ & \vdots & \\ \dot{v}_{\rho-1} & = & v_\rho \\ \dot{v}_\rho & = & \left(\tilde{v} - v - \sum_{\alpha=1}^{\rho-1} k_\alpha v_{\alpha+1}\right)\Big/ k_\rho \end{cases} \tag{2.77}$$

$$\dot{\eta}_s = q(\xi_s, \eta_s) \tag{2.78}$$

Observe that the dynamics of variables η_s does not affect the output, which evolves according to the selection of parameters k_α, with $1 \leq \alpha \leq \rho$. Indeed, (2.78) determines the non-linear hidden (and internal) dynamics, which must be stable for the whole system to be internally stable.

2.7   Robustness properties

The SMRC method proposed in this chapter inherits the robustness properties of SM control, which are studied here for a general non-linear system.

Suppose that the system (2.53) is affected by model uncertainties and external disturbances such that

$$\dot{x} = f(x) + g(x)u + d = (f(x) + \Delta f(x)) + g(x)u + \delta \tag{2.79}$$

where $d \in \mathbb{R}^n$ may represent both parametric uncertainty ($\Delta f(x)$) and external non-structured disturbances (δ). The disturbance vector d can be decomposed as

$$d = g(x)\mu(x) + \zeta(x) \tag{2.80}$$

where $\mu(x) : \mathbb{R}^n \to \mathbb{R}$ is a smooth scalar function, and thus the component $g(x)\mu(x)$ is collinear with $g(x)$, whereas $\zeta(x)$ is a vector that belongs to the manifold tangent to the surface $\mathcal{S}$. Observe that this decomposition can always be carried out since $g(x)$ is not tangent to the surface because of the SM transversality condition (2.61).

Let us analyse the effects of d on the existence domain and the reduced dynamics of sliding regimes.

2.7.1   SM existence domain

From (2.79) and the invariance conditions for the surface $\mathcal{S}$

$$\dot{\sigma}(x) = L_f\sigma + L_g\sigma u_{eq} + L_d\sigma = L_f\sigma + L_g\sigma u_{eq} + L_{g\mu+\zeta}\sigma = 0 \qquad (2.81)$$

and applying the linear properties of the directional derivative

$$\dot{\sigma}(x) = L_f\sigma + L_g\sigma u_{eq} + L_\zeta\sigma + L_g\sigma\mu = 0 \qquad (2.82)$$

Then, the equivalent control is given by

$$u_{eq}(x) = -\frac{L_f\sigma + L_\zeta\sigma + L_g\sigma\mu}{L_g\sigma} \qquad (2.83)$$

Since ζ is tangent to $\mathcal{S}$ by definition, $L_\zeta\sigma = 0$. Hence,

$$u_{eq}(x) = -\frac{L_f\sigma}{L_g\sigma} - \mu \qquad (2.84)$$

and calling u^*_{eq} to the equivalent control of the unperturbed system (see (2.55))

$$u_{eq}(x) = u^*_{eq} - \mu \qquad (2.85)$$

From this equivalent control, and according to (2.26), the necessary and sufficient condition for the SM establishment results

$$u^- < u^*_{eq} - \mu < u^+ \qquad (2.86)$$

$$u^- + \mu < u^*_{eq} < u^+ + \mu \qquad (2.87)$$

It is then concluded that the SM existence domain is not affected by the disturbance component $\zeta(x)$, which is tangent to the sliding surface, but it is indeed altered by the component $g(x)\mu(x)$, collinear with $g(x)$.

For SMRC schemes, this can be observed in the open-loop state equation of the conditioning loop (2.34), where we have $\mu(x) = \lambda_f r + \lambda_f d + \dot{d}$, $g(x) = b = [0 \ \lambda_f]^{\mathrm{T}}$ and $\zeta = 0$. It is evident that for SM existence, w must be large enough to compensate for bounded r, d and $\dot{d}$, which act as disturbances collinear with $g(x)$.

The above reasoning is geometrically illustrated by Figures 2.16 and 2.17. On the one hand, in Figure 2.16 the system is perturbed by a disturbance ζ tangent to the surface $\mathcal{S}$, and it can be observed that the value u_{eq} does not change with respect to the unperturbed one, which means that existence condition neither does it. On the other hand, Figure 2.17 shows how a perturbation d collinear with g affects the value of u_{eq} and thus the SM existence condition.

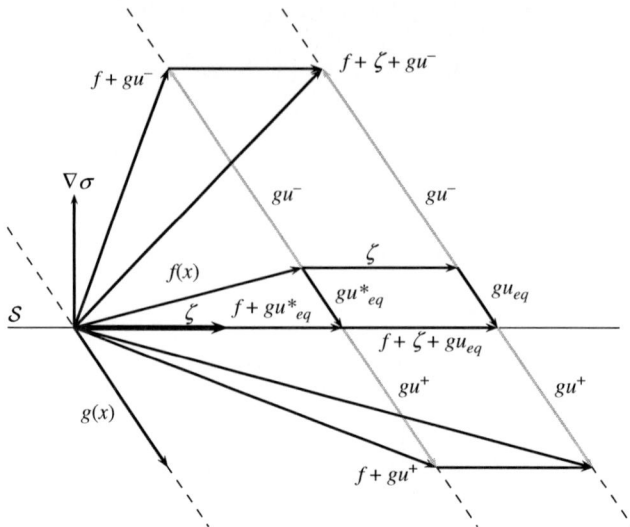

Figure 2.16   Geometric interpretation of the effects of a disturbance d tangent to surface S

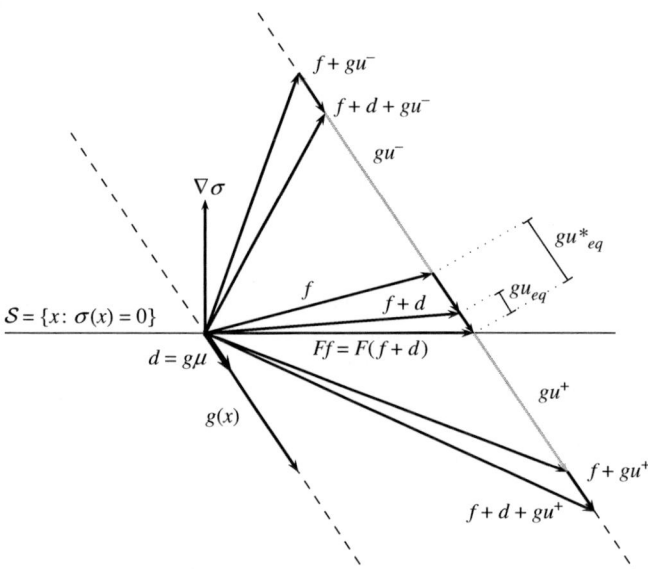

Figure 2.17   Geometric interpretation of the effects of a disturbance d collinear with the control field g

2.7.2 SM dynamics

To analyse the robustness of the SM dynamics against disturbances, the equivalent control is substituted in the state equation of the perturbed system ((2.79) and (2.80))

$$\dot{x} = f(x) + g(x)u_{eq}(x) + g(x)\mu(x) + \zeta(x) \tag{2.88}$$

From (2.85)

$$\dot{x} = f(x) + g(x)(u_{eq}^*(x) - \mu(x)) + g(x)\mu(x) + \zeta(x) \tag{2.89}$$

and thus

$$\dot{x} = f(x) + g(x)u_{eq}^*(x) + \zeta(x)$$

$$= F(x)f(x) + \zeta(x)$$

As can be seen, contrary to what happens with SM establishment, the SM dynamics is insensitive to the component $g(x)\mu(x)$, but it is affected by the term $\zeta(x)$ tangent to the surface S. This is also depicted by Figure 2.17 (projection of $F \cdot (f + d)$ coincides with the one of $F \cdot f$) and Figure 2.16 ($F \cdot (f + \zeta) = f + \zeta + gu_{eq}^*$ differs from $F \cdot f = f + gu_{eq}^*$), respectively.

In SMRC approaches, this general result is particularly verified by (2.36), (2.46), (2.77) and (2.78), which determine the reduced dynamics during SMRC. In fact, these equations depend neither on the reference input r nor on the disturbance d.

It is said that SM dynamics presents a *strong invariance* property to a perturbation d when its ideal sliding dynamics is completely independent of d [114]. From the above discussion, this is verified, provided the disturbance only presents non-zero component in the direction of the control field, i.e.

$$d = g(x)\mu(x) \tag{2.90}$$

This condition is known as *matching condition*, and was briefly introduced in Section 2.4.2. It is opportune recalling that SMRC equations (2.34), (2.42) and (2.45) satisfy the matching condition.

Remark 2.6: *The above analysis is of great importance to determine in each particular application against which kind of uncertainty or disturbances the SMRC approach is completely robust. This will be verified in the case studies of the next chapter and the multivariable problems addressed in Chapters 5–8, for which the robustness properties of SMRC algorithm will be further analysed.*

Chapter 3

Some practical case studies

In this chapter, the practical potentials of the sliding mode reference conditioning (SMRC) approach to deal with different kinds of constraints are illustrated with four case studies: (1) the pitch control of wind turbines with both amplitude and rate actuator saturation; (2) a clean hydrogen production system with structural constraints in which the electrolyser specifications require output bounds; (3) the tracking speed autoregulation of robotic manipulators in order to avoid path deviations; and (4) the regulation of ethanol concentration below a given threshold in the fed-batch fermentation of an industrial strain for overflow metabolism avoidance.

3.1 Pitch control in wind turbines

In this section, we evaluate the application of the SMRC technique presented in Chapter 2 to compensate for actuator constraints in the pitch control of a wind power system. Blade pitch rotation is the most popular technique used in wind turbines to control the aerodynamic torque under high wind speed conditions. Pitch actuators present a hard limit on their rate of change together with the natural amplitude saturation. This *physical limit* makes the addition of anti-windup (AW) algorithms necessary to avoid windup. Because of the slow dynamics of pitch actuators, conventional AW methods are not always able to completely compensate for actuator limitations, particularly when facing extreme wind conditions. The SMRC method is applied here as a simple way to avoid windup in a robust fashion. The effectiveness of the algorithm is verified by simulation of an autonomous wind energy conversion system for water pumping.

3.1.1 Brief introduction to the problem

Modern high-power wind turbines are equipped with variable-pitch wind rotors and power electronic devices managing totally or partially the generated power. With this equipment, both the aerodynamic and the generator torques can be controlled. Two control loops are typically implemented: in one loop the electronic converters are commanded to control the generated power, whereas in the other loop the blade pitch angles are adjusted to keep the rotating speed within safe limits.

At low wind speeds, a maximum power tracking algorithm is followed by the electronic converters to capture as much energy as possible from the wind.

Meanwhile, the pitch control loop is inactive. At high wind speeds, the available wind power exceeds the power rating of the wind generator. So, the electronic converters are now controlled to regulate electric power at its rated value. On the other hand, the aerodynamic efficiency of the wind turbine needs to be reduced in order to keep balance between input and output powers. This is achieved by the pitch control loop that regulates the rotational speed within safe limits. In this case study, we focus on the pitch control loop.

The purpose of the pitch control loop can be better explained with the help of Figure 3.1 that shows a simplified operating locus of wind turbines on the power-speed plane. The dashed lines depict the aerodynamic power vs. speed characteristic of the wind rotor for different wind speeds and optimum pitch angle. It is observed that these curves exhibit a maximum that moves rightwards as wind speed increases.

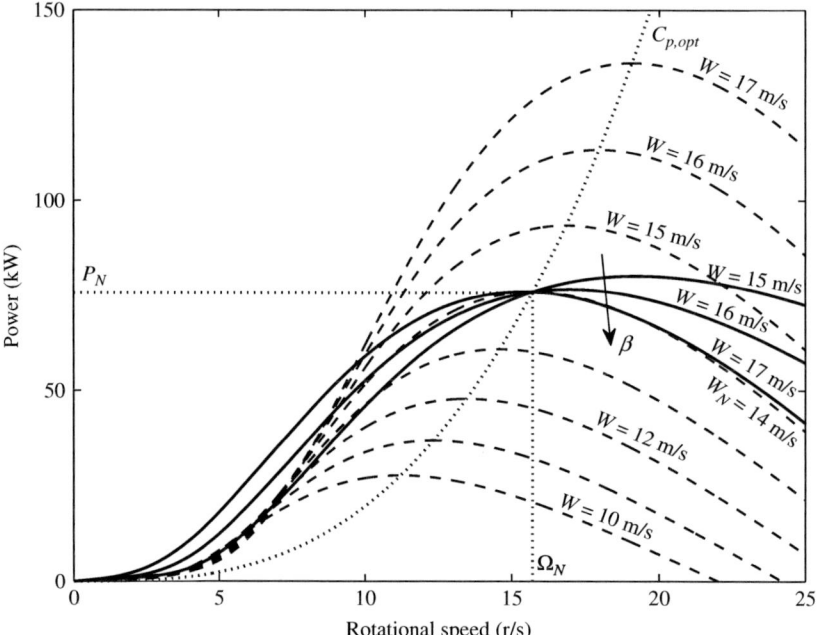

Figure 3.1   Wind turbine power vs. speed curves with (solid line) and without (dashed line) pitch control

The control strategy in the operating region of low wind speeds essentially consists of moving along the parabolic (dotted) line, labelled with $C_{p,opt}$, comprising all these maxima. This is the so-called maximum power point tracking (MPPT) strategy and is implemented by the electronic converter controller while the pitch angle is kept constant. In this mode of operation, the rotational speed varies in proportion to wind speed so that an optimum tip-speed ratio is maintained. At rated wind speed, the maximum power point coincides with the nominal operating point of the wind turbine. For higher wind speeds, the maximum power point cannot be further tracked, and a different control strategy is followed to avoid overloading. In this high wind speed range, the strategy consists of regulating the turbine at its nominal point. This requires shaping the wind rotor characteristic curves in such a way that they contain the nominal point. The curves in solid line show how this can be accomplished by properly pitching the blades.

In commercial wind turbines, the pitch controller is generally a gain-scheduled PI that regulates rotational speed at its rated value. As mentioned above, the pitch actuators are subjected to hard constraints that may cause controller windup. SMRC is evaluated here to avoid both amplitude and rate saturation of the pitch actuators, thus guaranteeing the closed-loop operation of the control system.

3.1.2 Pitch actuator and control

The pitch actuator is a hydraulic or electromechanical device that allows the rotation of the wind turbine blades around their longitudinal axes. It can be modelled as a first-order dynamic system with saturation in the output amplitude and rate of change. Figure 3.2 shows the block diagram of the actuator model, whose dynamic behaviour in its linear region is described by the differential equation

$$\dot{\beta} = -\frac{1}{\tau}\beta + \frac{1}{\tau}\beta_c \tag{3.1}$$

where τ is the actuator time constant, β is the pitch angle of the turbine blades and β_c is the controller output.

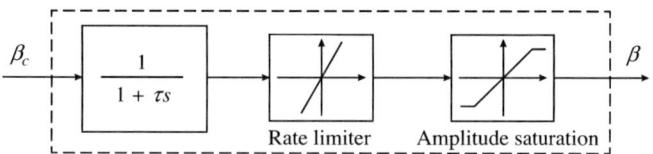

Figure 3.2 Pitch actuator model

When designing the control loop, it is of great importance to avoid a high activity of the blade pitch, since it could not only damage the pitch actuators but also give rise to unstable modes of operation if the actuator dynamics is not considered during controller design [10]. A priority objective is then to minimise the actuator activity so as to increment the structural robustness of the wind turbine, and to extend in this way the service life of the energy system. In addition, controller windup is prompt to occur because of the actuator speed and amplitude constraints. In fact, if the PI pitch controller requires a fast and/or excessive actuator action, the closed-loop operation can be lost, leading to undesired and dangerous transients [56,68].

3.1.3 SMRC compensation for actuator constraints in the pitch control loop

An SMRC scheme with AW features for the pitch controller is shown in Figure 3.3 [36]. Two blocks can be distinguished: the main pitch control loop based on the comparison of the rotational speed Ω_t with the rated speed Ω_N and an auxiliary compensation loop that, making use of the previous chapter ideas, shapes the reference speed Ω_{rf} in order to avoid actuator constraints. Observe that the wind speed acts as a disturbance for both control loops, since it directly affects the turbine speed. Also note that the SMRC loop deals with different types of constraints on β (namely, amplitude and rate saturation), and that it includes a model for the actuator dynamics since the signal β is not assumed to be accessible. In this regard, it is worth highlighting that this model should be chosen conservatively to account for uncertainty in such a way that its output reaches the limits before β.

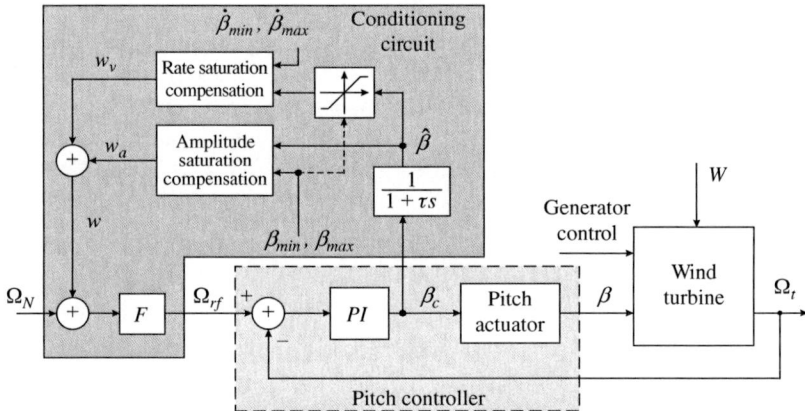

Figure 3.3   SMRC scheme for pitch control compensation

The following commutation law is implemented in each compensation block to avoid exceeding the pitch actuator limits:

$$w(t) = \begin{cases} -\Omega_N & \text{if} \quad \underline{\sigma}(t) > 0 \\ \Omega_N & \text{if} \quad \overline{\sigma}(t) < 0 \\ 0 & \text{otherwise} \end{cases} \tag{3.2}$$

Because of the non-negligible pitch actuator dynamics, whose model is included in the auxiliary loop, the constrained system is strictly proper. That is, recalling the terminology used in Chapter 2, the subsystem $S_c(s)$ has relative degree equal to one. Thus, according to Section 2.5, the commutation functions have to be defined as follows:

- For compensation of amplitude saturation

$$\underline{\sigma}(t) = \underline{\sigma}_a(t) = \beta_{min} - \hat{\beta} - k\dot{\hat{\beta}}$$
$$\overline{\sigma}(t) = \overline{\sigma}_a(t) = \beta_{max} - \hat{\beta} - k\dot{\hat{\beta}} \tag{3.3}$$

- For compensation of rate saturation

$$\underline{\sigma}(t) = \underline{\sigma}_v(t) = \dot{\beta}_{min} - \dot{\hat{\beta}}$$
$$\overline{\sigma}(t) = \overline{\sigma}_v(t) = \dot{\beta}_{max} - \dot{\hat{\beta}}$$

(3.4)

with β_{min} and β_{max} being the maximum pitch angles in both directions, and $\dot{\beta}_{min}$, $\dot{\beta}_{max}$ the corresponding rate bounds.

In this manner, the discontinuous auxiliary signal $w(t)$ is generated from (3.3) and (3.4) by the switching law (3.2), and then smoothed out by means of the first-order filter F in order to shape the conditioned speed reference Ω_{rf}.

Note that in (3.3), in addition to the pitch angle amplitude, the approaching speed of $\hat{\beta}$ to its limit value is also taken into account to decide the commutation. Considering, for example, the upper limit in the second line of (3.3), during the sliding regime the invariance condition will be verified and therefore $\overline{\sigma}_a(t) \equiv 0$. Hence, while the compensation loop is active

$$\dot{\hat{\beta}} = \frac{1}{k}(\beta_{max} - \hat{\beta})$$

(3.5)

Equation (3.5) shows that $\hat{\beta}$, and therefore β, tends to its limit amplitude with a dynamics imposed by the sliding surface as seen in Chapter 2. As k decreases, the approaching speed of β to β_{max} increases. Observe also from (3.5) that this dynamics is independent of rotational speed or wind speed variations.

3.1.4 Application to a wind energy system for water pumping

3.1.4.1 System description

The compensation scheme shown in Figure 3.3 is applied to an autonomous wind energy conversion system for water pumping. This system consists of a wind rotor driving a brushless doubly fed generator (BDFG) [104], electrically coupled to a water pumping unit. A schematic diagram of the installation is shown in Figure 3.4. A three-bladed wind turbine of 6 m radius is considered. A gearbox of ratio 1:2 is

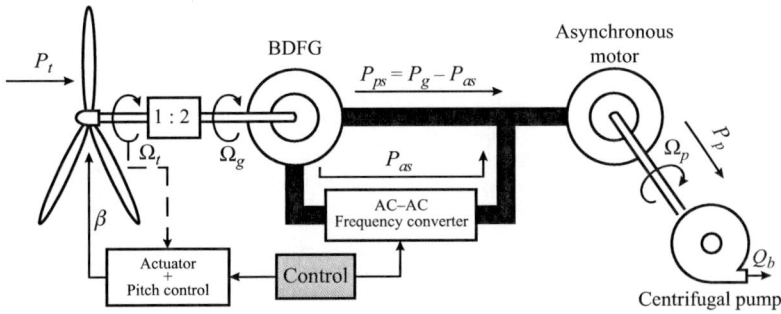

Figure 3.4   Autonomous wind energy conversion system for water pumping

interfaced between the wind rotor and the BDFG to match speeds. The pumping unit consists of a centrifugal water pump propelled by an induction motor with two pole pairs.

The power flow between the turbine and the pump can be controlled through the auxiliary stator of the generator, by managing the input voltage and frequency [15]. A generator torque control loop modifies the frequency by means of an AC/AC frequency converter that decouples the shaft speeds of turbine and pump. This control scheme allows implementing MPPT strategies for wind speeds below rated. At higher wind speeds, the pitch control loop is activated to keep the turbine operating at its nominal point.

The aerodynamic power captured by the wind turbine is given by the expression [14]

$$P_t = \frac{1}{2} A \rho_{air} C_p(\lambda, \beta) W^3 \tag{3.6}$$

where ρ_{air} is the air density, A is the blade swept area, W is the wind speed and C_p is the power efficiency of the wind rotor. C_p is function of the blades' pitch angle and the tip-speed ratio $\lambda = r\Omega_t/W$, with r being the blades' length. This efficiency function exhibits a maximum $C_{p,opt}$ at $(\lambda_{opt}, \beta_{opt})$ and is approximated here by the following mathematical expression [58]:

$$C_p(\lambda, \beta) = c_1 \left(\frac{c_2}{\lambda_i} - c_3\beta - c_4 \right) e^{(-c_5/\lambda_i)} + c_6\lambda \tag{3.7}$$

with

$$\frac{1}{\lambda_i} = \frac{1}{\lambda + 0.08\beta} - \frac{0.035}{\beta^3 + 1} \tag{3.8}$$

and c_1 to c_6 characteristic coefficients of the turbine.

The electric power P_g of the BDFG is divided into the main and auxiliary stator windings:

$$P_g = P_{ps} + P_{as} \tag{3.9}$$

where P_{ps} and P_{as} are the main and auxiliary stator powers, respectively. The power flow through the auxiliary stator, and hence the generator electrical power, is controlled by the AC/AC converter.

Finally, the behaviour of the water pumping unit can be described by its power-speed characteristic [138]:

$$P_p = k_p \cdot \Omega_p^3 \tag{3.10}$$

with k_p being the pump constant and Ω_p the pump shaft speed, which is governed by the frequency of the main stator voltage.

The dynamic behaviour of the whole system can be modelled by the first-order differential equation

$$\frac{d\Omega_t}{dt} = \frac{1}{J\Omega_t}(P_t - P_p) \qquad (3.11)$$

where J is the inertia of the drive train referred to its low-speed side.

3.1.4.2   Simulation results

It is recalled first that the objective of the pitch control loop is to regulate the turbine speed at its rated value during operation under high wind speed conditions. In this context, the purpose of the SMRC algorithm is to avoid the undesirable effects of pitch actuator constraints on the drive-train dynamics. For the simulation tests, realistic limit values on the amplitude and slew rate of the pitch actuator were considered: $\beta_{min} = 0°$, $\beta_{max} = 20°$ and $\dot{\beta}_{max} = -\dot{\beta}_{min} = 10°/s$.

Figure 3.5 shows the results obtained by the pitch control system with (solid line) and without (dashed line) the proposed compensation technique when facing wind speed steps that surpass the rated wind speed (Figure 3.5(a)). Although such fast variations in wind speed do not occur in the real world, the response to wind speed steps was tested to illustrate the SMRC performance under severe wind gusts.

As can be appreciated in Figure 3.5(b), the SMRC loop effectively avoids overloading the pitch actuator, protecting it from consecutive saturations and control action fluctuations. Observe that in the pitch control loop without the adaptive compensation the actuator slew rate breaks the closed-loop operation, making the controller output (dashdotted) inconsistent with the blade pitch angle β (dashed). Conversely, in the compensated loop β_c and β only slightly differ due to the pitch actuator first-order dynamics (hardly noticeable in the figure), avoiding in this way the actuator saturation. Figure 3.5(c) reveals that the improvement in the signal transmitted to the pitch actuator is due to the establishment of transient sliding regimes on the surfaces $\overline{\sigma}_v(t) = 0$ and $\underline{\sigma}_v(t) = 0$, which shape the reference speed Ω_{rf} so that the controller output matches the physical actuator limits (Figure 3.5(d)).

Finally, the boxes (e) and (f) in Figure 3.5 show the time evolution of the aerodynamic and pump powers, respectively, whereas Figure 3.6 displays the corresponding power-speed curves up to $t = 15\,s$. They verify that the SMRC strategy reduces the oscillations around the nominal speed and power with a much less aggressive control action.

3.2   Clean hydrogen production plant

This case study regards a wind–hydrogen energy conversion system. Modern configurations of this sort of system tend to reduce hardware components. This results in a *structural constraint*, which requires matching the wind power output to the electrolyser power requirements. It is shown here that this objective can be

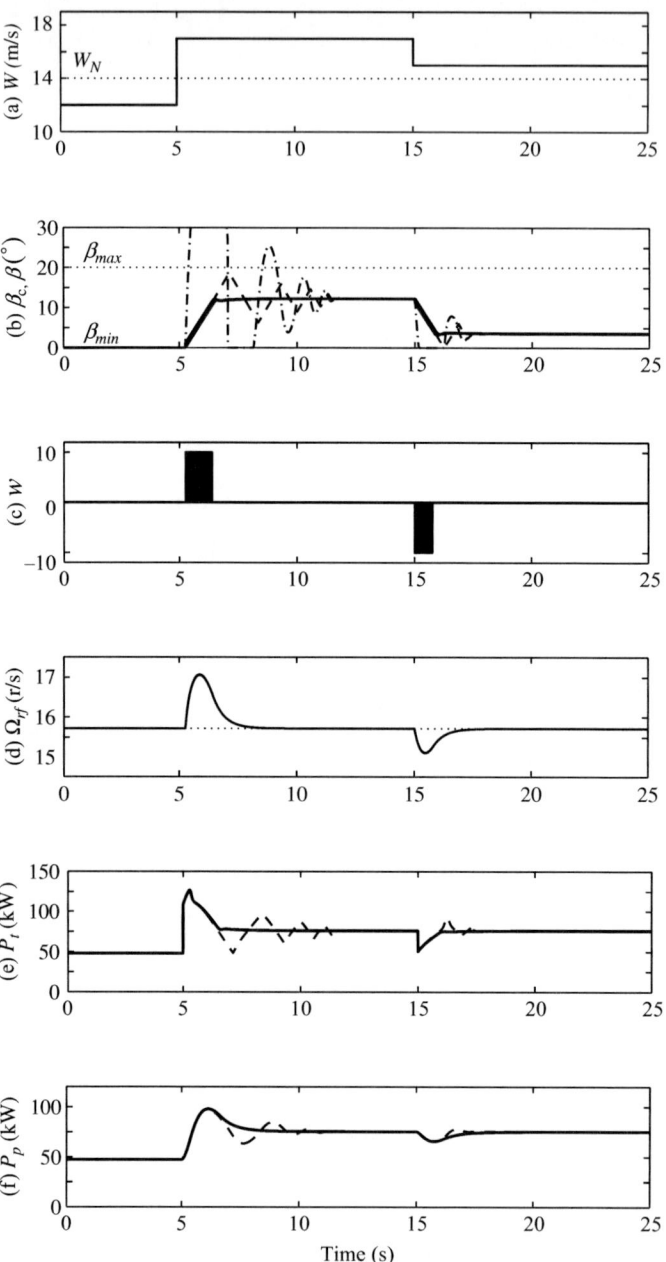

Figure 3.5   Simulation results for sudden wind speed changes with (solid) and without (dashed) AW compensation

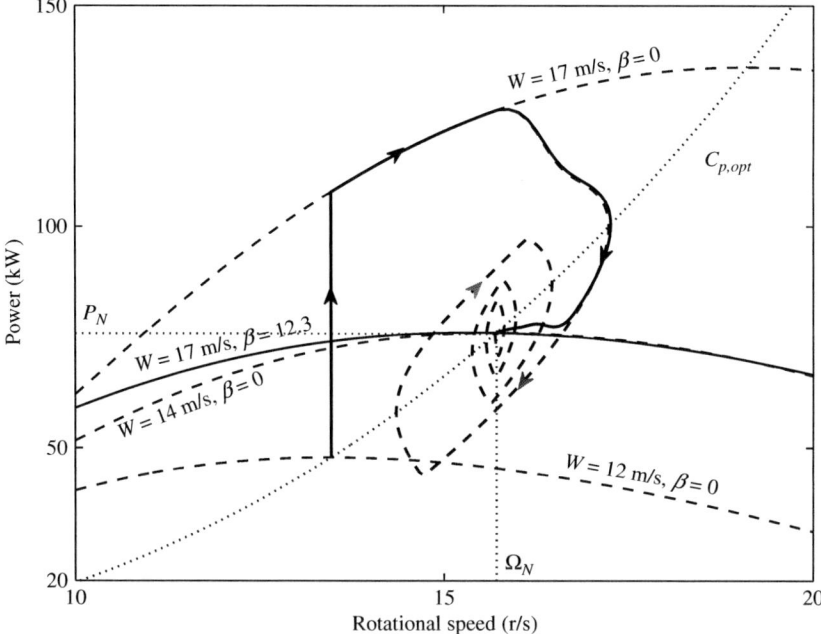

Figure 3.6   Power vs. rotational speed curve of the turbine with (solid) and without (dashed) AW compensation corresponding to Figure 3.5 for $t \leq 15$ s

achieved by continuously shaping the power reference of an MPPT algorithm using reference conditioning ideas. This allows attaining high aerodynamic power conversion efficiency while simultaneously fulfilling the electrolyser specifications, which are actually output bounds for the wind turbine control system.

3.2.1   Brief introduction to the problem

An increasingly valued alternative to deal with the intermittence and seasonal variability of clean energy primary resources is the production of hydrogen through water electrolysis. Hydrogen can then be subsequently used either to produce electricity or to supply fuel engines, and it also allows both storage and transportation of large amounts of energy at much higher densities (energy/volume) than traditional storage options, such as battery-based systems [21,111,139].

Because of the comparatively low cost of wind technology, along with the persistent growth in installed wind power capacity over the world, wind electrolysis is the favourite candidate to become the first economically viable renewable hydrogen production system. Two different approaches are found with regards to the configuration of stand-alone wind-electrolysis system.

In many cases, the turbine and electrolyser, with their own dedicated power electronics and controllers, are connected to a constant-voltage DC-bus. Such a decoupled configuration presents the advantage that wind turbine and electrolyser can be controlled separately. In this scheme, the AC voltage at the wind-driven generator terminals is converted to a constant DC-bus voltage through an AC–DC converter, whereas a DC–DC converter takes the DC-bus voltage and provides a suitable DC voltage for electrolyser operation.

The other approach tries to reduce hardware complexity, for instance by eliminating component duplication in the interface between the turbine and the electrolyser. In fact, the pair of power converters and the DC-bus can be replaced with a single AC–DC converter, taking the AC generator voltage and providing a suitable DC voltage to the electrolyser. Thus, the efficiency of the energy conversion can be increased and the overall cost of the installation can be appreciably reduced [27,31]. Obviously, this coupled configuration requires the development of controllers specifically designed for this application.

Grid-connected wind turbines are typically controlled to maximise the power capture within their safe limits. However, in stand-alone applications, this conventional control strategy may be in conflict with the load requirements. This is the case when a wind turbine is used to supply an electrolyser with its own specifications. Hence, the control approach followed here is to shape the reference of a conventional MPPT algorithm in such a way that the output power is compatible with the electrolyser requirements. The resulting control scheme is based on the SMRC algorithm discussed in Chapter 2.

3.2.2   System description

3.2.2.1   The plant

The wind-electrolysis system under consideration is sketched in Figure 3.7. Its main components are

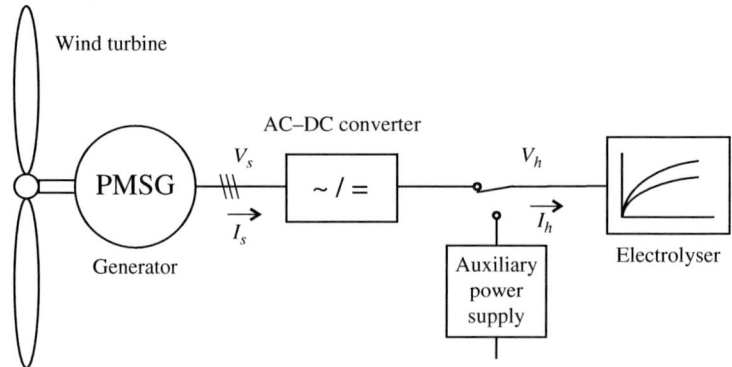

Figure 3.7   Wind-powered hydrogen plant scheme

- *Wind rotor.* The essential concepts of wind turbine aerodynamics have been introduced in the previous case study. In this case, to emphasise possible conflicts between wind turbine and electrolyser control objectives, we will

consider operation under low wind speed conditions. With the aim of max-
imising energy capture in low winds, wind turbines are controlled to operate at
their maximum conversion efficiency, that is with optimum tip-speed ratio λ_{opt}
and optimum pitch angle (say, $\beta = 0$). This MPPT algorithm entails variable-
speed fixed-pitch operation. Figure 3.8 depicts the typical shape of the power
coefficient C_p as function of the tip-speed ratio λ showing a maximum $C_{p,opt}$
at λ_{opt}.

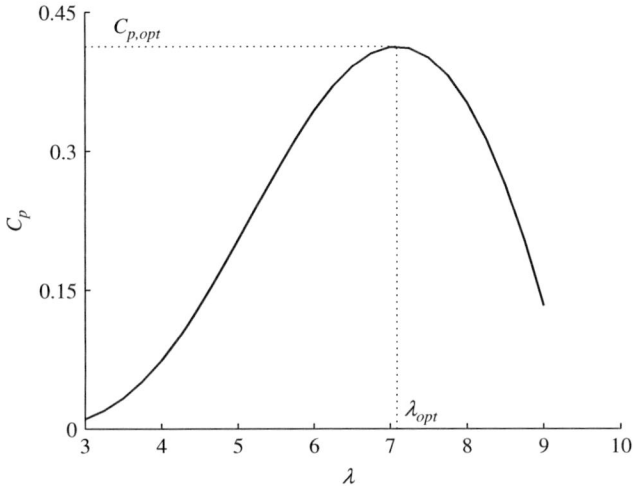

Figure 3.8   Power coefficient of the turbine as function of the tip-speed ratio for
$\beta = 0$

- *Permanent magnet synchronous generator (PMSG)*. A simple model of this
  generator is given by a three-phase sinusoidal voltage source E_f in series with a
  synchronous reactance X_S. Voltage V_s at the generator terminals slightly differs
  in magnitude and phase from the emf E_f because of the voltage drop across X_S
  when the generator current I_s passes through it. The mechanical and electrical
  variables of the generator are related as follows:

$$E_f = \frac{p}{2}\Phi\Omega_t$$
$$P_g = 3I_sV_s \cos\phi$$

(3.12)

where p is the number of poles of the machine, Φ is the stator flux linkage, P_g
is the generator power and ϕ is the phase angle between V_s and I_s [87].
- *Electronic converter*. The electronic devices are aimed at converting the AC
  three-phase speed-dependent generator voltage V_s into a DC voltage V_h suitable for hydrogen production. As was previously mentioned, a single AC–DC
  converter is considered as an interface between the turbine and the electrolyser.

The output and input voltages of the converter are related by the controlled factor δ, which is associated with the duty cycle of the electronic switching devices. Since incoming and outgoing powers are matched, i.e. $P_g = P_h$, input and output currents have a similar ratio. That is,

$$
\begin{aligned}
\delta V_s &= V_h \\
\delta I_h &= 3 I_s \cos \phi
\end{aligned}
\tag{3.13}
$$

We assume $\cos \phi = 1$, i.e. the input voltage and current are in phase. This is the case in many AC–DC converters, including those having an uncontrolled rectifier at the input [87]. Note that the duty cycle modifies the impedance seen from the AC side of the converter, and thereby the generator load, according to

$$
\frac{V_s}{I_s} = 3\delta^{-2} \frac{V_h}{I_h}
\tag{3.14}
$$

- *Electrolyser.* The electrolyser is modelled by its current–voltage $(I_h - V_h)$ curve, approximated here by the logarithmic law [132]:

$$
V_h = V_r + n(T)I_h + m(T) \log\left(1 + \frac{q(T)}{T^2} I_h \right)
\tag{3.15}
$$

where V_r is the reversible cell voltage and $n(T)$, $m(T)$ and $q(T)$ are quadratic polynomials in temperature T. Figure 3.9 plots approximately the $I_h - V_h$ characteristic of a 2.25 kW von Hoerner electrolyser for an operating temperature $T = 39°C$ [109].

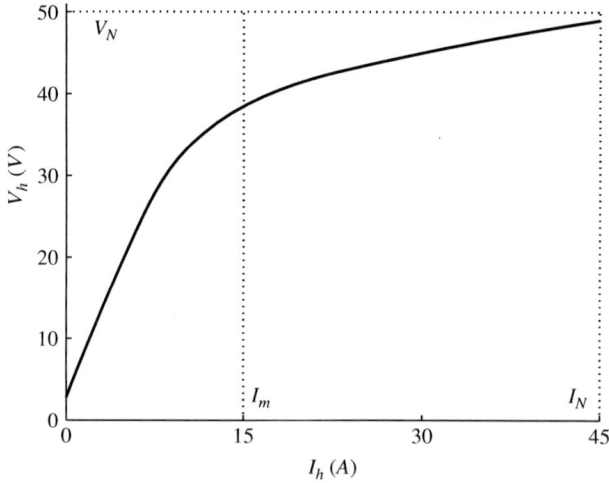

Figure 3.9   Current–voltage characteristic of the electrolyser

3.2.2.2 Conventional MPPT control scheme

Figure 3.10 shows a schematic diagram of the control system. The plant comprises the wind rotor, the generator, the AC–DC electronic converter and the electrolyser represented by their static characteristics of (3.6), (3.12), (3.13) and (3.15), whereas the dynamic behaviour is governed by

$$\frac{d\Omega_t}{dt} = \frac{1}{J\Omega_t}(P_t - P_h) \tag{3.16}$$

where J is the inertia of the drive train. In Figure 3.10, two blocks representing the MPPT controller and the reference conditioning algorithm are highlighted. Let us ignore for the moment the conditioning circuit and focus on the MPPT, i.e. let us suppose that $P_{rf} = P_r$, with P_r being the power reference to be tracked. It is generated from Ω_t as

$$P_r = \frac{1}{2}\rho_{air}Ar^3 \frac{C_{p,opt}}{\lambda_{opt}^3} \Omega_t^3 \tag{3.17}$$

This expression represents the maximum power locus in the operating region of the wind turbine (i.e. the curve $C_{p,opt}$ in the power-speed plane in Figure 3.1), and is obtained from P_t in (3.6) for optimum tip-speed ratio. This reference power is compared with the output power P_h. The power error is the input to a PI controller that commands the duty cycle of the converter. As a result, the turbine speed Ω_t converges to its optimum value $\Omega_{opt} = \lambda_{opt}W/r$. Hence, the maximum power is successfully tracked with a second-order closed-loop dynamics. However, since they have not been taken into consideration yet, this MPPT algorithm may violate the constraints imposed by the electrolyser specifications. The second step in the controller design is precisely to cope with these limitations.

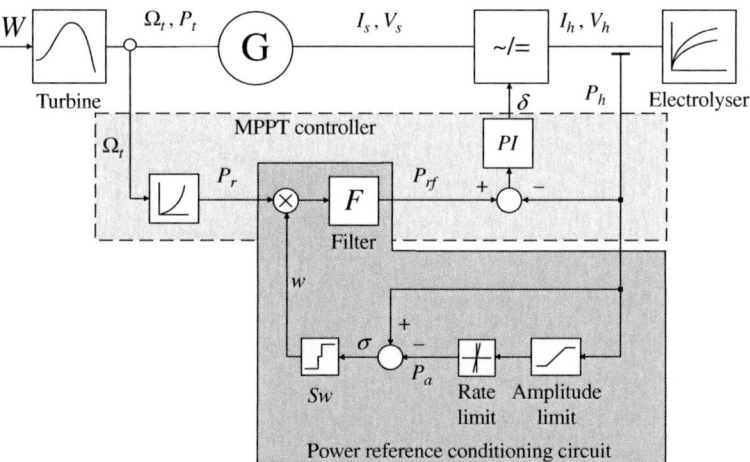

Figure 3.10 Scheme of the control system with the proposed power conditioning circuit

3.2.3   SMRC algorithm to deal with electrolyser constraints

The proper operation of the electrolyser imposes several constraints on its power supply [24]:

1. Operation at rated current levels, as close as possible to the rated value I_N (see Figure 3.9), is expected since efficiency increases in proportion to electric current.
2. A minimum current I_m should be guaranteed, with minimal connection to the auxiliary power supply, otherwise the produced hydrogen quality will be poor.
3. The power supply should be smoothed since current fluctuations may increase the internal wear as well as the impurities and energy losses.

These specifications of the electrolyser translate into amplitude and rate constraints on the controlled variable, i.e. the electrolyser power P_h. To satisfy these constraints, an SMRC algorithm can be added to shape the reference P_{rf} of the MPPT control algorithm [20]. Note that P_h and P_{rf} are the output and input, respectively, of a cascade system comprising the PI controller and the AC–DC converter. Since both of them have null relative degree, the constrained subsystem ($S_c(s)$ in Chapter 2) is biproper. Therefore, the SMRC can be designed following the guidelines presented in Sections 2.3 and 2.4.

The first stage of the SMRC circuit generates an admissible power $P_a(t)$ by passing the actual electrical power P_h through an amplitude saturation block and a rate-limiting block. This admissible power ranges over $\underline{P_a}(t) \leq P_a(t) \leq \overline{P_a}(t)$ at any time t, where the upper-limit signal $\overline{P_a}(t)$ is determined by I_N or the upper power rate limit, and the lower-limit signal $\underline{P_a}(t)$ is determined by I_m or the lower power rate limit. So, the specifications for the electrolyser will be effectively satisfied whenever the actual electrolyser power equals this admissible power ($P_h(t) = P_a(t)$). Then, there is a switching block that generates a discontinuous signal w as function of the mismatches between real and admissible powers. This discontinuous signal is used to shape the reference of the MPPT controller through the first-order filter F:

$$F : F(s) = \frac{1}{1 + s\tau_F} \tag{3.18}$$

On the basis of this analysis, the following switching function is defined:

$$\sigma = P_h - P_a(t) \tag{3.19}$$

See that $\sigma \equiv 0$ determines a region in the state space where P_h is constrained within its admissible limit values $\underline{P_a}(t)$ and $\overline{P_a}(t)$. In fact, this region defined by $\mathcal{R} = \{\sigma = P_h - P_a(t) = 0\}$ is delimited by two time-varying surfaces $\overline{S}$ and $\underline{S}$ defined by $\overline{S} = \{\overline{\sigma} = P_h - \overline{P_a}(t) = 0\}$ and $\underline{S} = \{\underline{\sigma} = P_h - \underline{P_a}(t) = 0\}$, respectively. Note that, in contrast with the previous case study, we can define here a single switching function to compensate for both amplitude and rate constraints. This is a characteristic of biproper conditioned systems.

Then, for SM establishment on the borders of the desired operating region $\mathcal{R}$, the following switching law is proposed:

$$w : \begin{cases} w^- = 1 & \text{if} \quad \sigma < 0 \\ w^0 = k & \text{if} \quad \sigma = 0 \\ w^+ = 0 & \text{if} \quad \sigma > 0 \end{cases} \tag{3.20}$$

Naturally, the sustainable limits for the conditioning action are optimum power ($w \equiv 1$) and no power ($w \equiv 0$). Signals w^- and w^+ are designed in consequence. Then, $w^0 = k < 1$ is designed as a compromise between efficiency and available control effort. With regards to energy production, the parameter k should be selected close to 1 so that $P_{rf} \cong P_r$ during desired operation mode ($\sigma \equiv 0$). Nevertheless, a factor k close to 1 would minimise the capability of conditioning the electrolyser power during sudden wind speed decrease. It must be remarked that this type of compromise commonly appears in control of wind energy systems when the quality of the power supplied to the grid or load is an important issue. In our application, sacrificing some percentage of the wind energy may be largely rewarded by the improvement of the electrolyser power quality as well as of the produced hydrogen one. This is particularly true if one considers that the turbine is often oversized to avoid multiple connections and disconnections of the electrolyser to the turbine.

Note that in wind energy systems, the available control effort to attenuate the effects of wind drops is bounded. In the current application, this means that the SM conditioning cannot be sustained under drastic wind reductions, in which case the backup power supply should be connected. This limitation is not attributable to the control strategy but to the wind resource availability.

3.2.4 Simulation results

The simulation results presented in this section were obtained for a system like the one shown in Figure 3.7 with a 2.25 kW electrolyser and a 5 kW wind turbine. The wind speed profile used in simulations is plotted in Figure 3.11.

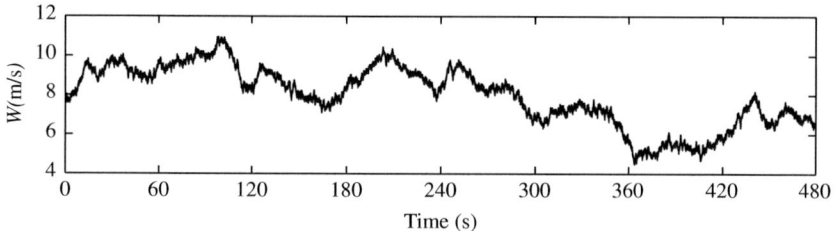

Figure 3.11 Wind speed profile

3.2.4.1 Maximum power tracking control

The results obtained with a conventional MPPT algorithm are shown in Figure 3.12. Figure 3.12(a) depicts the power captured by the turbine and supplied to the

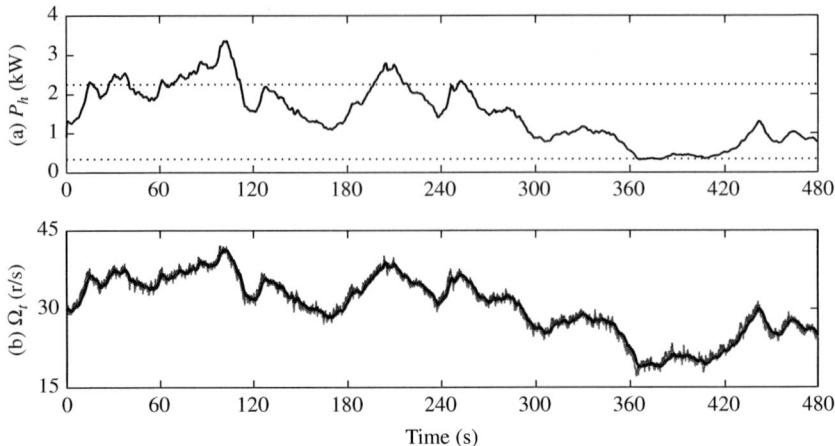

Figure 3.12   Simulation results with MPPT. (a) Electrolyser power. (b) Actual (thick trace) and optimum (noisy trace, overlapped) turbine speed

electrolyser, whereas Figure 3.12(b) shows the optimum speed profile overlapped by the actual turbine speed. The optimum speed (for maximum power capture) is successfully tracked, and the electrolyser power also overlaps the optimum turbine power. Note that although the optimum operating point of the turbine is correctly tracked, the MPPT performance does not accomplish the power requirements of the electrolyser. In fact, the supplied power reproduces the fast wind variations, exceeding during some time intervals its upper limit (2.25 kW), falling below its lower limit (0.35 kW) and presenting excessively fast variations as well.

3.2.4.2   Power conditioning via SMRC

The simulation results obtained with the SMRC algorithm are shown in Figure 3.13. The top graph (a) plots the power supplied to the electrolyser P_h, which tightly follows the conditioned reference P_{rf}, graph (b) exhibits the turbine speed response, graph (c) displays the discontinuous conditioning signal and the bottom box (d) depicts the sliding function. As can be seen, the electrolyser power has been bounded and smoothed, thereby improving the quality of the power supplied to the electrolyser.

As already mentioned, in this case the main control basically comprises a conventional MPPT algorithm where the reference power locus is obtained by multiplying $P_r(\Omega)$ by a coefficient $k < 1$, particularly $k = 0.7$ $\left(P_{rf} = 0.7P_r(\Omega)\right)$. Note that this selection of k does not imply a 30% reduction of the captured power. Actually, the system searches a new operating point at higher speeds, for which the considered system loses around 7% of the energy [20].

Figure 3.13(c) reveals that the simulation run alternates intervals of SM conditioning with periods of no correction. The SM conditioning periods are identified by the fast switching behaviour of the discontinuous signal w. Moreover, it can be verified how sliding regimes occur on the upper-limit surface $\bar{\sigma} = 0$ when w

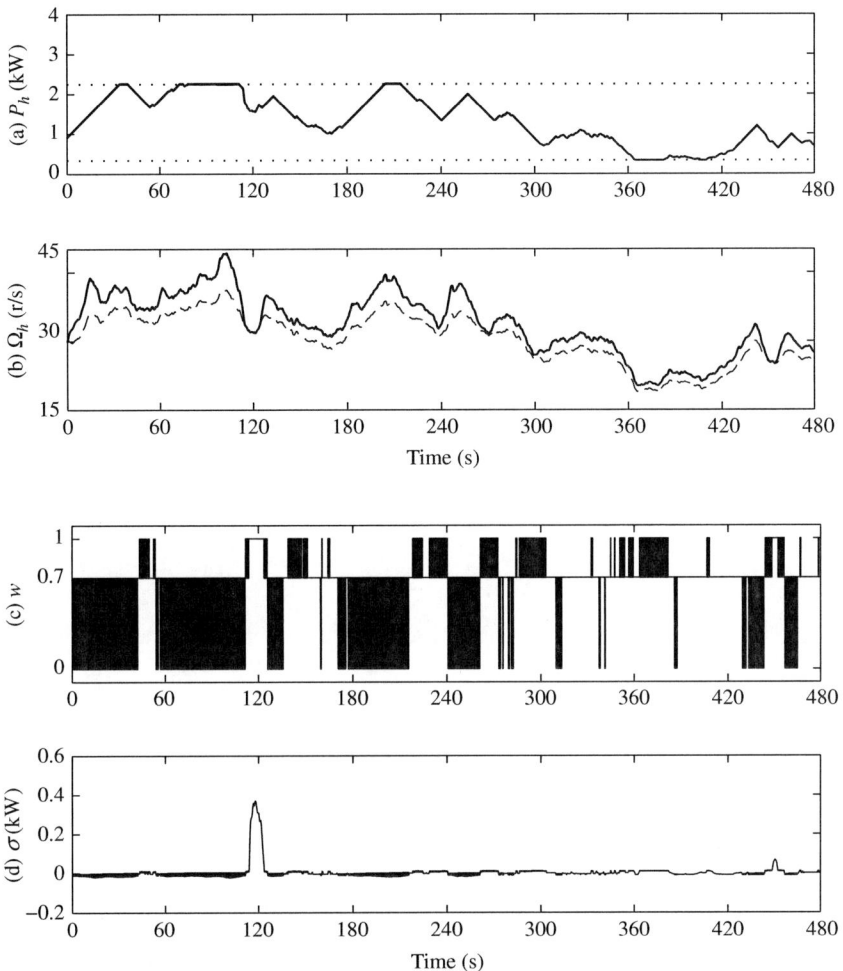

Figure 3.13   Simulation results with MPPT and reference conditioning

switches between w^0 and 0, whereas they are established on the lower-limit surface $\underline{\sigma} = 0$ when w switches between w^0 and 1. The conditioning loop is inactive during the time intervals where w is fixed at w^0.

Figure 3.13(a) shows how the SMRC algorithm shapes the electrolyser power with the aim of accomplishing the specifications. For instance, from $t = 0$ s to $t = 34$ s, the sliding regime is able to constrain P_h to its maximum permissible rate of change despite the rapidly increasing wind power. Similarly, between $t = 40$ s and $t = 54$ s the sliding regime limits the decreasing rate of change to its lowest admissible value. The SM conditioning between $t = 78$ s and $t = 114$ s successfully confines I_h (and hence P_h) to its rated value. Also, from $t = 365$ s to $t = 382$ s approximately, the SM increases the reference power, thus avoiding I_h to fall below I_m. During other remaining periods, the electrolyser power evolves smoothly and within its permissible values without SM conditioning. See, for instance, the response between $t = 380$ s and $t = 430$ s. It is worthy to mention that a turbulent wind speed profile was used in simulations to assess the performance of the constrained control algorithm under severe wind conditions. This is the reason why the conditioning algorithm is active most of the time during the simulation run.

Figure 3.13(b) illustrates how the turbine adjusts its speed in order to supply an appropriate amount of power to the electrolyser. Note that the speed is more than the optimum turbine speed most of the time, particularly during regulation at rated power or maximum power rate, whereas it approaches Ω_{opt} during regulation at minimum power or minimum power rate.

Figure 3.13(d) confirms that P_h is maintained within the desired region of operation ($\sigma \equiv 0$) all the time except for a short period around $t = 120$ s. During this period, characterised by a fast decreasing gust, the SMRC loop applies the maximum control effort to increase as much as possible the captured power. Thus, during this period the controller behaves as an MPPT control and the turbine speed coincides with Ω_{opt}.

3.3   Robot path tracking

SMRC ideas are herein employed to develop a simple method that allows automatic regulation of the robotic tracking speed in order to avoid path deviations caused by (multiple) joint actuator constraints. A sufficient condition for SM establishment on the limiting surfaces is derived in terms of the commanded tracking speed. This non-linear application of SMRC is illustrated for a classical 2R manipulator under kinematic control.

3.3.1   Brief introduction to the problem

In most practical applications using industrial and/or mobile robots [99,112] (e.g. machining, arc welding, assembling, inspection, transport, adhesive application, spray painting, etc.), the robot task is to track a given *path* as accurately and fast as possible. In this manner, both quality and productivity indexes can be enlarged. However, the accuracy and the speed with which this tracking can be performed are strongly related to the *physical limitations* of joint actuators, which are seldom considered in commercial robots to regulate the robot forward motion. Instead, the tracking speed usually has to be computed a priori by the robot operator in order to avoid an error message.

Basically, the following three approaches can be found in practical applications in order to face with robot actuator power constraints:

1. To use a (conservative) low tracking speed, so that the robot control signals never exceed their maximum values.
2. To also use a fixed tracking speed, but higher than the previous one, in such a way that the robot control signals saturate at least once during the tracking.
3. To compute for each point on the path the maximum tracking speed allowed by the limits of the control signals.

The first approach is extremely conservative and thus not an advisable solution. In effect, it gives rise to an excessively slow path tracking, which wastes the tracking capabilities of the robotic system. The second approach, which is the classical one, has a main drawback: when the control signals are saturated, the robot loses the reference and even leaves the desired path, which makes it inappropriate for high-accuracy applications. The third option is the best choice among the three listed. However, it depends on the desired path and on the robot Jacobian, thus being computationally more involved and non-robust against modelling errors or disturbances.

Next, we explore the application of SMRC concepts to address the above robotic tracking problem, without requiring a priori knowledge of the desired path and independently of the main trajectory tracking control and the Jacobian computation.

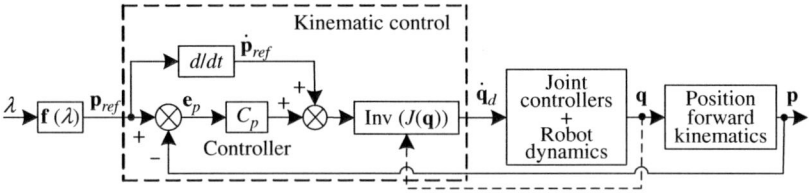

Figure 3.14 Robotic path-tracking control scheme

3.3.2 Classical control scheme for robotic path tracking

In most practical robot systems, the controller consists of three nested control loops: an analog actuator current controller, an analog velocity controller and a typically digital position controller. The great majority of industrial robot manufacturers implement the inner control loops (i.e. the current loop and the velocity loop) internally in the so-called *joint controllers* and do not allow the robot operator to modify these loops. Conversely, the outer-loop position controller is usually open for the user and can be manipulated.

Figure 3.14 shows the classical control for the position or kinematic loop, which consists of a correction of the position error $\mathbf{e}_p$ by means of the controller C_p plus the feedforward of the first-order time derivative of the position reference, i.e. $\dot{\mathbf{p}}_{ref}$.[1] The position reference $\mathbf{p}_{ref}$ (desired path) can be expressed in terms of a function $\mathbf{f}(\lambda)$ whose argument is the so-called motion parameter $\lambda(t)$ as

$$\mathbf{p}_{ref} = \mathbf{f}(\lambda) \tag{3.21}$$

and therefore

$$\dot{\mathbf{p}}_{ref} = \frac{\partial \mathbf{f}}{\partial \lambda} \dot{\lambda} \tag{3.22}$$

---

[1] Throughout this subsection, we will denote vectors with boldface type since it is standard notation in robotics.

Note that in this two-degrees-of-freedom scheme, the error correction is performed in the workspace coordinates, and then the inverse of the robot Jacobian J is used to obtain the joint velocity vector $\dot{\mathbf{q}}$. Indeed, for a non-redundant manipulator, the relationship between the $\mathbf{q}$ configuration and the end-effector position/orientation $\mathbf{p}$ is highly non-linear and expressed as

$$\mathbf{p} = \mathbf{l}(\mathbf{q}) \qquad (3.23)$$

where the function $\mathbf{l}$ is called the kinematic function of the robot model. The first-order kinematics results in

$$\dot{\mathbf{p}} = \frac{\partial \mathbf{l}}{\partial \mathbf{q}} \dot{\mathbf{q}} = J(\mathbf{q})\dot{\mathbf{q}} \qquad (3.24)$$

where $J(\mathbf{q})$ is denoted as the Jacobian matrix or simply *Jacobian* of the kinematic function. Consequently, the joint velocities $\dot{\mathbf{q}}$ producing a particular end-effector motion $\dot{\mathbf{p}}_0$ can be written as

$$\dot{\mathbf{q}} = J^{-1}(\mathbf{q})\dot{\mathbf{p}}_0 \qquad (3.25)$$

and the kinematic control loop is in charge of determining the desired value for $\dot{\mathbf{p}}_0$ as a function of current position $(\mathbf{p})$, target trajectory point $(\mathbf{p}_{ref})$ and speed $\dot{\mathbf{p}}_{ref}$. Once $\dot{\mathbf{p}}_0$ is computed, (3.25) is applied and sent to the actuators.[2]

Another conventional approach for kinematic control consists of performing the error correction directly in the joint space. In any case, the SMRC technique also applies for that or any other kinematic control.

Now, joint speed saturation is considered between the desired joint speed $\dot{\mathbf{q}}_d$ shown in Figure 3.14 and the achievable one $\dot{\mathbf{q}}_{ds}$. Such an actuator limitation is typically faced in two different ways:

1. As a direct and *independent saturation* element for each joint, for which

$$\dot{q}_{ds,i} = \begin{cases} \dot{q}_{max,i} & \text{if } \dot{q}_{d,i} > \dot{q}_{max,i} \\ \dot{q}_{d,i} & \text{if } \dot{q}_{min,i} \leq \dot{q}_{d,i} \leq \dot{q}_{max,i}, \quad i = 1, \ldots, n \\ \dot{q}_{min,i} & \text{if } \dot{q}_{d,i} < \dot{q}_{min,i} \end{cases} \qquad (3.26)$$

   with $\dot{q}_{max,i}$ and $\dot{q}_{min,i}$ denoting the maximum and minimum actuator (speed-servo) output of the corresponding joint, and n the robot degree-of-freedom order. For the sake of simplicity, it is assumed in the following: $\dot{q}_{min,i} = -\dot{q}_{max,i}$.

2. As a *directionality-preserving saturation*, in which the joint speeds vector $(\dot{\mathbf{q}}_{ds})$ direction remains constant, hence the direction of $\dot{\mathbf{p}}$ is also preserved, but its modulus is scaled as

$$\dot{\mathbf{q}}_{ds} = f_{dir} \cdot \dot{\mathbf{q}}_d \qquad (3.27)$$

---

[2] For a guide to the computation of the Jacobian for different robot types see, for instance, Reference 2.

where

$$f_{dir} = \begin{cases} \frac{1}{\|F_N \cdot \dot{\mathbf{q}}_d\|_\infty} & \text{if} \quad \|F_N \cdot \dot{\mathbf{q}}_d\|_\infty \geq 1 \\ 1 & \text{if} \quad \|F_N \cdot \dot{\mathbf{q}}_d\|_\infty < 1 \end{cases} \tag{3.28}$$

notation $\| \cdot \|_\infty$ denotes element-wise maximum (infinity norm) and

$$F_N := diag\left(\frac{1}{\dot{q}_{max,i}}\right) \tag{3.29}$$

The latter is the most frequent approach in robotic tracking schemes to address joint speed limitations, and thus it will be the case considered in the SMRC auto-regulation algorithm.

3.3.3   Tracking speed autoregulation technique

The application of SMRC ideas to regulate the path following speed is sketched in the block diagram in Figure 3.15. It assumes that the input to the path-tracking system is the desired path speed $\dot{\lambda}_d$. Such input signal will be 'conditioned' in such a way that saturation is avoided. Note from Figure 3.15 that the constrained sub-system (S_c in Chapter 2) is in this case highly non-linear.

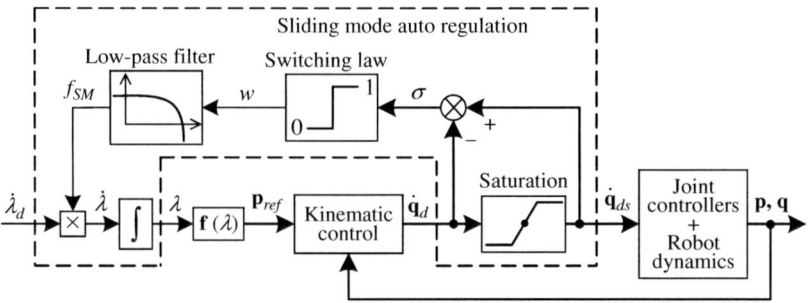

Figure 3.15   Proposed sliding mode autoregulation technique for robotic path tracking

Since the SMRC autoregulation technique is thought to be used with com-mercial industrial robots, a kinematic scheme as the one in Figure 3.14 is con-sidered in the 'kinematic control' block of Figure 3.15. The saturation block, as already mentioned, assumes *directionality-preserving* saturation. Nevertheless, any other actuator non-linearity, time dependency, asymmetry, etc. could be transpar-ently considered by suitably changing the saturation block.

To generate the conditioned tracking speed, an auxiliary SMRC loop is added to the conventional path-tracking configuration, in which the discontinuous signal w is determined by means of

$$w = \begin{cases} w^+ = 1 & \text{if} \quad \sigma(\mathbf{x}) = \mathbf{0}_n \\ w^- = 0 & \text{if} \quad \sigma(\mathbf{x}) \neq \mathbf{0}_n \end{cases} \tag{3.30}$$

where $\mathbf{0}_n$ denotes the null vector of dimensions $n \times 1$, and the switching function vector σ is defined as

$$\sigma(x) = \dot{\mathbf{q}}_{ds} - \dot{\mathbf{q}}_d \tag{3.31}$$

The maximum trajectory speed $\dot{\lambda}$, which avoids inconsistencies between the kinematic controller outputs and the real robot inputs, is then generated from w by means of the filter

$$F : \begin{cases} \dot{x}_f = a_f x_f + b_f w \\ f_{SM} = c_f x_f \end{cases} \tag{3.32}$$

with the filter output f_{SM} being a 'sliding mode factor' so that $\dot{\lambda}$ is obtained as $\dot{\lambda} = f_{SM} \cdot \dot{\lambda}_d$. The filter dynamics given by $a_f = -c_f b_f < 0$ must have unit DC-gain and should be chosen sufficiently fast for quick stops to be allowed but slow enough to smooth out $\dot{\lambda}$.

Similar to what happened in the previous applications, the switching function vector (3.31) determines in this case two boundary-sliding surfaces for each one of the n coordinates

$$\begin{aligned} \overline{S}_i &= \{\mathbf{x} | \dot{q}_{d,i} - \dot{q}_{max,i} = 0\} \\ \underline{S}_i &= \{\mathbf{x} | \dot{q}_{d,i} - \dot{q}_{min,i} = 0\} \end{aligned} \quad i = 1, \ldots, n \tag{3.33}$$

From (3.22), (3.25), (3.27)–(3.29) and (3.32), it can be easily deduced that these surfaces are only reached provided the desired speed satisfies

$$\dot{\lambda}_d \geq \frac{1}{\| F_N \cdot J^{-1} \frac{\partial \mathbf{f}}{\partial \lambda} \|_\infty} \tag{3.34}$$

in at least one point of the path. This condition is equivalent to first line of the necessary and sufficient condition for sliding mode establishment (2.13).

Observe that if $\dot{\lambda}_d$ verifies (3.34) on *all points of the trajectory*, at least one of the actuators will be saturating at maximum speed and, hence, the SMRC algorithm implicitly provides the maximum-speed directionality-preserving trajectory tracking solution. However, this reaching condition is not enforced by the switching action. Indeed, (3.34) shows how the maximal achievable tracking speed depends on the own path reference $\mathbf{f}$, the robot kinematics J, the actual robot configurations (J is function of $\mathbf{q}$) and, naturally, the saturation levels F_N. In fact, (3.34) is particularly small for robot configurations close to singularities or against abrupt trajectory changes. In the former case, some singular values of J^{-1} are very big; in the latter case, it is $\partial \mathbf{f}/\partial \lambda$ that takes a very large value.

Here, as the forward evolution on the trajectory reference is limited by actuator i, a sliding regime will transiently establish on either the surface $\overline{S}_i$ or $\underline{S}_i$. If because of the desired trajectory the robot is forced to saturate other joint actuator, say actuator $j \neq i$, then SM will be reached on surfaces $\overline{S}_j$ or $\underline{S}_j$, one of which will be from then onwards responsible for the path following speed attenuation.

For the trivial sliding function vector (3.31) to satisfy the unitary relative degree necessary condition, apart from the filter being a first-order one, the controller C_p must be biproper. For the case of (very unusual) strictly proper controllers, the strategy is still applicable, but the sliding function should be redefined as seen in Section 2.5.

Effect of the feedback term. The proposed kinematic control has two components: a feedback error-based control and a feedforward trajectory generator (2-DOF structure). It is easy to understand that, through the feedback term, the SMRC technique for motion-parameter conditioning stops the set-point movement if initial conditions or transient tracking error are big. The intensity of this effect depends on the gain of the feedback regulator: the larger the regulator gain, the smaller the error bound, which stops the motion parameter ($\dot{\lambda} = 0$). This is in agreement with what it is expected from conventional feedback regulators (the larger the gains, the smaller the errors). Hence, the feedback-gain tuning conveniently tunes both the closed-loop behaviour regarding disturbance rejection and modelling error and the regulation of the path speed reference if errors are high.

Effect of the robot Jacobian. Note that the present SMRC tracking speed autoregulation does *not* require any Jacobian computation, as it is independent of the underlying kinematic control loop. This is indeed one of its main distinctive advantages.

Effect of chattering. Recall that SMRC is confined to the low-power side of the system, where fast electronic devices can be used to implement the discontinuous action. Thus, differing from conventional SM approaches, the present application does not impose any bound to the switching frequency f_s in order to protect the robot actuators and/or mechanisms, significantly reducing in this way the problems related with chattering.

In any case, assuming a given sample time T_s, the chattering amplitude $\Delta\sigma_i$ so that the condition $|\sigma_i| \leq \Delta\sigma_i$ is guaranteed during SM can be bounded by

$$\Delta\sigma_i \leq |\dot{\sigma}_i| T_s \leq \left| [J^{-1}(\mathbf{q})]_{i,*} \frac{\partial f}{\partial \lambda} a_f \dot{\lambda}_d \right| T_s \tag{3.35}$$

with $[J^{-1}(\mathbf{q})]_{i,*}$ the ith row of the inverse Jacobian. Therefore, chattering can be reduced if necessary by either decreasing the filter bandwidth or increasing the sampling rate.

3.3.4 Application to a 2R manipulator

The SMRC algorithm for autoregulation of the tracking speed is illustrated in this subsection by its application to the classical two bar mechanism with two planar revolute joints, where the two linear positions of the end effector are considered.

Although it is a simple mechanism, the 2R manipulator is commonly employed to show the effectiveness of novel control or tracking strategies since it captures the main phenomenon of robotic path tracking. The application of the SMRC speed auto-regulation to the well-known 6R robotic arm PUMA-560 can be found in Reference 37.

Figure 3.16 depicts the schematic diagram of a 2R manipulator, from which it is straightforward that its kinematic function is

$$\mathbf{p} = \mathbf{l}(\mathbf{q}) \rightarrow \begin{bmatrix} x \\ y \end{bmatrix} = L \begin{bmatrix} \cos(\theta_1) + \cos(\theta_1 + \theta_2) \\ \sin(\theta_1) + \sin(\theta_1 + \theta_2) \end{bmatrix} \tag{3.36}$$

while its first-order kinematics $\dot{\mathbf{p}} = J\dot{\mathbf{q}}$ results

$$\begin{bmatrix} \dot{x} \\ \dot{y} \end{bmatrix} = L \begin{bmatrix} -\sin(\theta_1) - \sin(\theta_1 + \theta_2) & -\sin(\theta_1 + \theta_2) \\ \cos(\theta_1) + \cos(\theta_1 + \theta_2) & \cos(\theta_1 + \theta_2) \end{bmatrix} \begin{bmatrix} \dot{\theta}_1 \\ \dot{\theta}_2 \end{bmatrix}$$

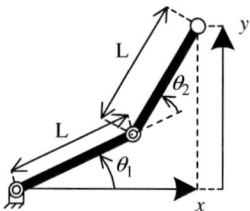

Figure 3.16   2R manipulator diagram

Four cases have been considered in the simulations – Case 1: joint speeds $\dot{\mathbf{q}}_{ds}$ ideally unconstrained, tracking speed $\dot{\lambda}$ constant; Case 2: joint speeds subjected to direct saturation, $\dot{\lambda}$ constant; Case 3: joint speeds subjected to directionality-preserving saturation, $\dot{\lambda}$ constant; and Case 4: joint speeds subjected to directionality-preserving saturation, $\dot{\lambda}$ autoregulated with the SMRC technique.

A sinusoidal reference was taken as the desired path to be followed. Its equations are given by

$$\begin{bmatrix} x_{ref} \\ y_{ref} \end{bmatrix} = \begin{bmatrix} (2 - \lambda)L \\ (c_y + c_a \sin(n_c\pi(\lambda - c_x)/(2 - c_x)))L \end{bmatrix} \tag{3.37}$$

with motion parameter $\lambda \in [c_x, (4 - c_x)]$, where c_x and c_y are phase and vertical coordinate offset parameters and c_a is the sinusoid amplitude. For the sake of simplicity, a proportional controller has been used for the correction of the position error, i.e. $C_p = K_p$.

Simulations were carried out from an off-track initial end-effector position $(x(0), y(0)) = (1.8, 0.2)$, while the reference path with $L = 1$, $c_x = 0.3$, $c_y = 0.3$, $c_a = 0.1$ and $n_c = 2$ starts at $(x_{ref}(0), y_{ref}(0)) = (1.7, 0.3)$. The actuator bounds were set as $\dot{q}_{max,i} = -\dot{q}_{min,i} = 1.5$ rad/s and the controller gain as $K_p = 10$.

Figure 3.17 depicts the target path (solid line) and the actually followed path for constant-speed classical approaches. The unconstrained case (thin solid line) is

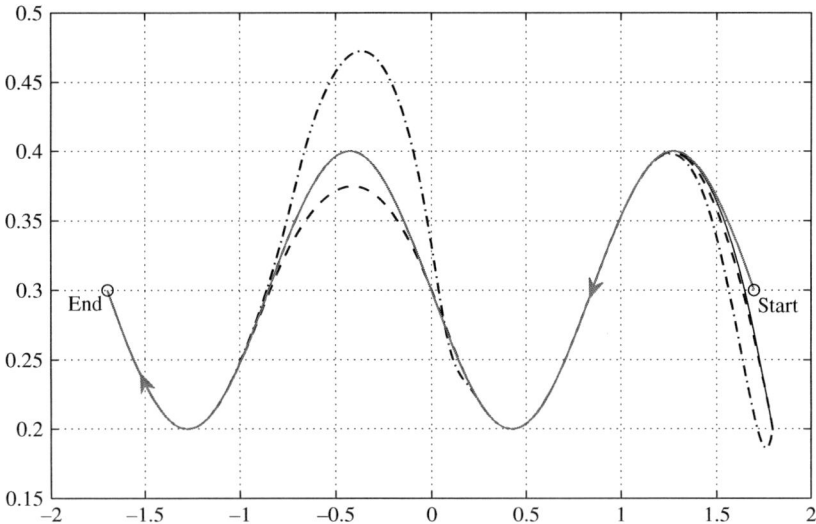

Figure 3.17 Desired path vs. followed paths (from right to left) with conventional approaches

also plotted for comparative purposes. As can be appreciated, both the direct (dashdotted) and the directionality-preserving (dashed) saturation approaches are not able to follow the given reference for the speed chosen ($\dot{\lambda} = 0.68$). They would require 'hand-tuning' λ to achieve the goal of fast-tracking without drifting out of the desired path. This is in fact what the proposed autoregulation performs automatically.

The path followed when using the SMRC autoregulation technique is contrasted with the target path in Figure 3.18. The proposed algorithm not only attains a tight tracking of the reference path but also covers it much faster. This is evident from Figures 3.19 and 3.20, where the joint velocities and position errors (time evolutions) of the first three cases are compared with the ones obtained with SM conditioning. Although conventional approaches take 5 s to cover the two sinusoid periods, the latter approach takes only 3.74 s. The reason can be found indeed in Figure 3.19: differing from classical approaches, for the desired motion-parameter speed the SMRC technique does always enforce a joint speed to reach its maximum value, thus minimising the task required time. Naturally, conventional approaches can be sped up by increasing λ, but at the expense of greater position errors. The autoregulation of the tracking speed λ and the resulting motion parameter are shown in Figure 3.21. The signal λ evidences some remaining chattering, which depends on the chosen tracking speed, the filter bandwidth (given by a_f) and the simulation step T_s. For this example, $a_f = 50$ and $T_s = 0.5$ ms were employed. Slower filter bandwidth or faster sampling rates would reduce the chattering amplitude.

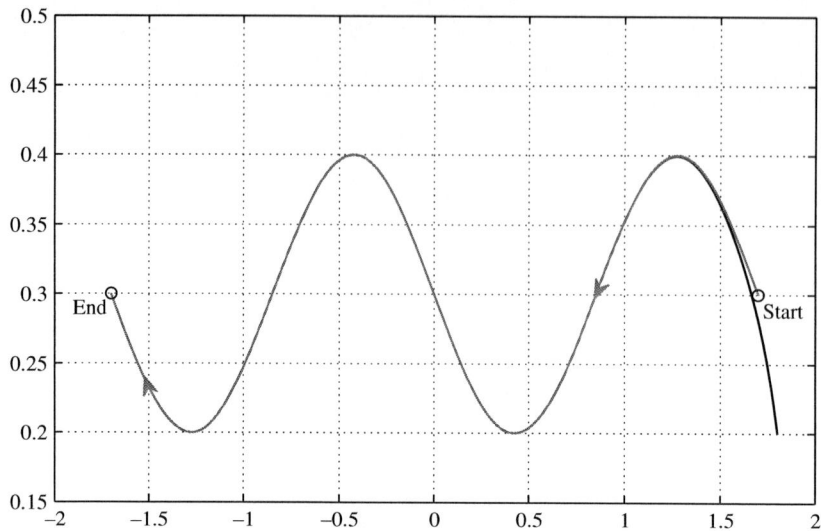

Figure 3.18   Desired path vs. followed path (from right to left) with proposed autoregulation technique

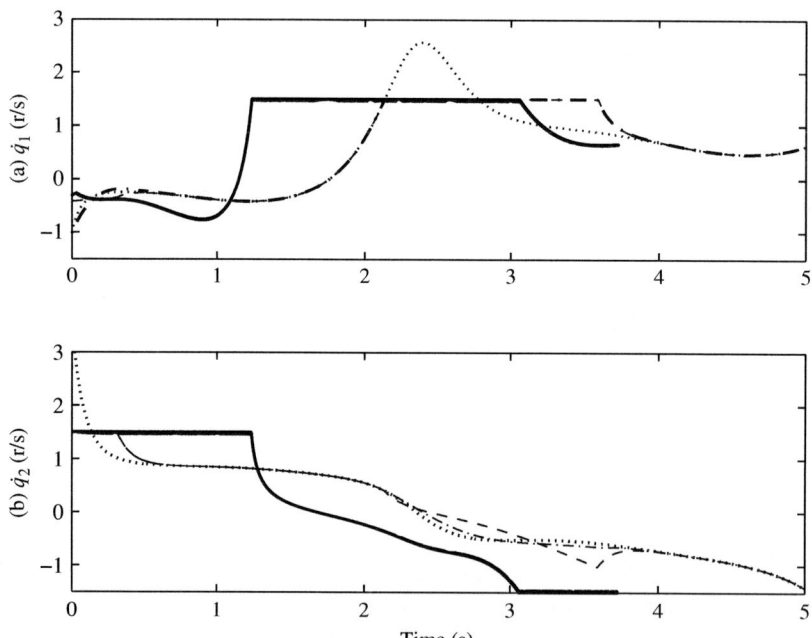

Figure 3.19   Joint velocities q_d for ideal unconstrained case (dotted), direct saturation (dashed-dotted), directionality-preserving saturation (dashed) and autoregulation technique (solid)

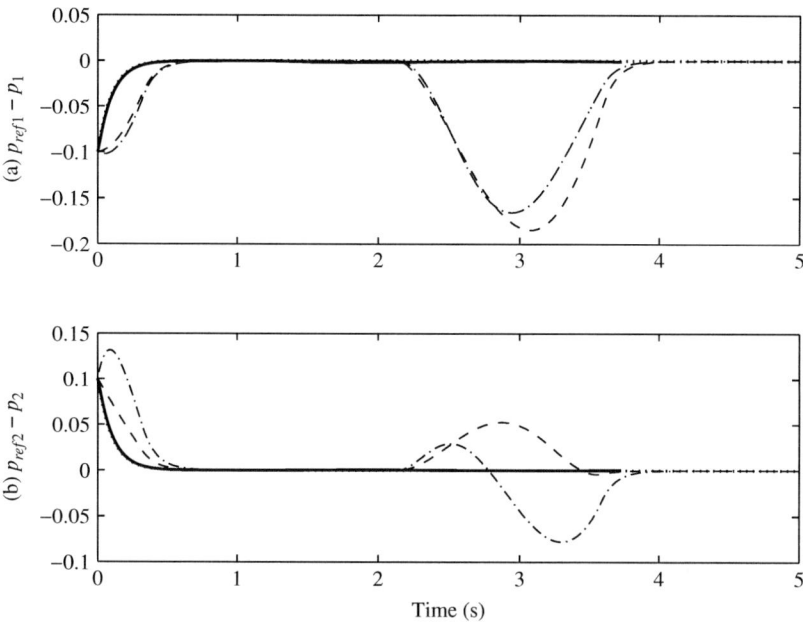

Figure 3.20 Position errors for ideal unconstrained case (dotted), direct saturation (dashed-dotted), directionality-preserving saturation (dashed) and autoregulation technique (solid)

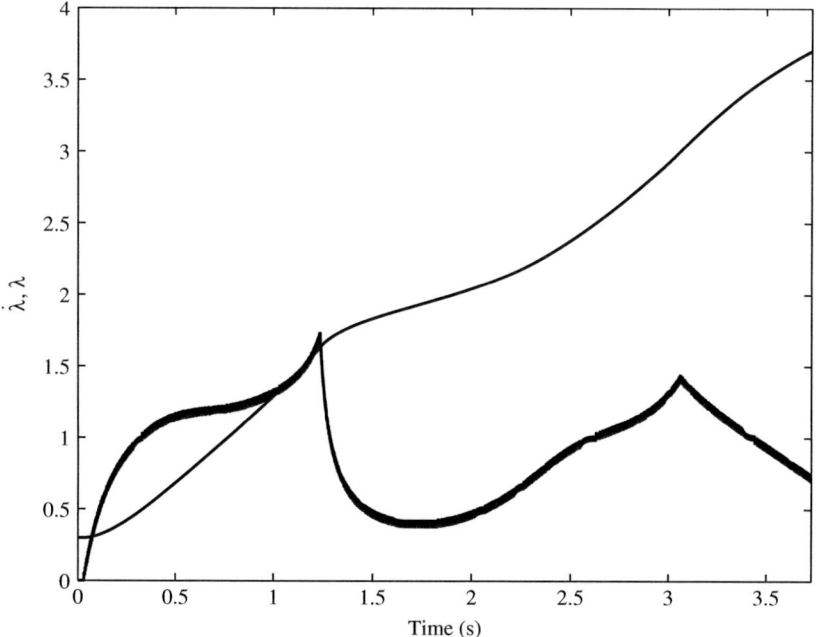

Figure 3.21 Tracking speed $\dot{\lambda}$ and motion parameter λ generated by the autoregulation technique

3.4   Control of a fed-batch bioreactor

In this last case study, we illustrate the application of SMRC principles to deal with constraints in a highly non-linear and uncertain process: the fed-batch fermentation of the industrial strain *Saccharomyces cerevisiae*. We also analyse for this complex control problem the robustness properties of the SMRC scheme and the resulting internal dynamics during SM operation. Further details on this biotechnological application are presented in Reference 97.

3.4.1   *Brief introduction to the problem*

In many fermentation processes, optimal productivity corresponds to operating at critical substrate concentration, which can be regulated by controlling the feed rate. Typically, this critical substrate value changes from experiment to experiment and from strain to strain, and even in the same experiment due to changing environmental and/or process conditions. For some bioprocesses, the optimum operating point might correspond to the maximum of some kinetic function. In such a case, extremum seeking strategies can be used [130]. Yet, if *dynamic restrictions* are present – e.g. due to the production of additional toxic or inhibitory metabolites – the optimal point may not correspond to a maximum of the kinetic rates. This is the case, for instance, when dealing with biomass production of *S. cerevisiae*. The optimum is attained for a feeding profile that avoids the production of ethanol and is well below the maximum attainable specific growth rate (see Reference 122).

This situation is caused by the so-called overflow metabolism, a metabolic phenomenon occurring when the oxidative capacity of the cells is exceeded, resulting in by-product formation [101]. Thus, ethanol is produced in several yeast species [122], acetate in *Escherichia coli* [152,158] and lactate in mammalian cells [140].

Overflow metabolism appears at too high growth rates. Therefore, controlling the substrate feed rate is important to achieve high cell concentrations while avoiding overflow metabolism and/or accumulation of toxic substrates [59]. In this way, the problem consists of finding the feeding rate, which gives the closest specific growth rate to the desired one and which is compatible with the critical constraint (e.g. ethanol concentration lower than a given threshold), so as to avoid overflow metabolism. A methodology to cope with this problem that uses SMRC ideas is presented next.

3.4.2   *Process model*

The model of *S. cerevisiae* metabolism based on the bottleneck hypothesis of Reference 122 has been extensively used for control purposes. Its main traits are briefly summarised in the following text. The mass balance macroscopic model is given by

$$
\begin{aligned}
\dot{x} &= (\gamma_1 r_1 + \gamma_2 r_2 + \gamma_3 r_3)x - x\frac{E}{v} \\
\dot{s} &= -(r_1 + r_2)x + (s_i - s)\frac{E}{v} \\
\dot{e} &= (\gamma_4 r_2 - r_3)x - e\frac{E}{v} \\
\dot{v} &= F
\end{aligned}
\tag{3.38}
$$

where x, s and e are the concentrations of biomass, substrate and ethanol in the broth, v is the volume and F the inlet flow. Biomass is produced through three paths: oxidation of glucose, reduction of glucose and oxidation of ethanol, represented by the kinetic terms r_1, r_2 and r_3, respectively. Oxidation of glucose takes place in presence of glucose and oxygen. If there is an excess of glucose over the oxidative capacity of the cell, this will overflow into the second metabolic path, and glucose will be reduced, resulting in ethanol production. On the other hand, if the oxidative capacity of the cell is not fully used up by the glucose, the remaining capacity can be used to oxidise ethanol through the third metabolic path. This behaviour is modelled by means of the following reaction rates and kinetic terms:

$$r_1 = \min\left(r_s, \frac{r_o}{\gamma_5}\right)$$
$$r_2 = \max\left(0, r_s - \frac{r_o}{\gamma_5}\right) \tag{3.39}$$
$$r_3 = \max\left(0, \min\left(r_e, \frac{r_o - \gamma_5 r_s}{\gamma_6}\right)\right)$$

$$r_s = \mu_{m,s}\frac{s}{k_{s,s}+s}$$
$$r_e = \mu_{m,e}\frac{e}{k_{s,e}+e} \tag{3.40}$$
$$r_o = \mu_{m,o}\frac{o}{k_{s,o}+o}$$

where o denotes the oxygen concentration. Indeed, notice from the second and third equations in (3.39) that ethanol is only produced if there is an excess of substrate (glucose), so that $r_s > r_o/\gamma_5$ (see r_2), and ethanol is oxidised if there is enough oxygen, so that $r_o > \gamma_5 r_s$ (see r_3). Table 3.1 shows typical values of the model parameters.

Table 3.1 Typical parameter values for S. cerevisiae

Yield coefficients					
γ_1	γ_2	γ_3	γ_4	γ_5	γ_6
0.49	0.05–0.12	0.5–1.2	0.48	0.396	1.104
Kinetic parameters					
$\mu_{m,s}$	$k_{s,s}$	$\mu_{m,e}$	$k_{s,e}$	$\mu_{m,o}$	$k_{s,o}$
3.5	0.01	0.17	0.1–0.5	0.256	0.1

Values taken from References 122 and 137.

Using standard notation and considering an exponential feeding law $F = \lambda xv$, with the gain λ to be determined by the control loop, the model (3.38) can be written as

$$\Sigma: \begin{cases} \dot{x} = (\mu_x - \lambda x)x \\ \dot{s} = -\mu_s x + \lambda(s_i - s)x \\ \dot{e} = (\mu_e - \lambda e)x \end{cases} \tag{3.41}$$

where the volume dynamics has not been included for the sake of simplicity. The model given by (3.41) will be used in the following text for constrained control purposes.

3.4.3 Reference seeking for overflow avoidance

Differently from the approaches used in References 137, 103 and 23 where the goal is to regulate ethanol at a given low set point, the goal here is to control the biomass-specific growth rate $\mu_x = \gamma_1 r_1 + \gamma_2 r_2 + \gamma_3 r_3$ in (3.38) to its maximum value so that the ethanol concentration e lies below a given threshold. Note that this desired reference for the biomass-specific growth rate is unknown and possibly time varying (due, for example, to variations in the available oxygen). Assuming the desired (conditioned) reference is obtained by an outer SMRC loop to be designed below, it can be reached by means of any of the dynamic controllers proposed in the literature. See, for instance, References 8, 98 or 120.

To exemplify the SMRC algorithm operation, the following controller is implemented:

$$\lambda = \lambda_r[1 - k_c(\mu_r - \hat{\mu}_x)] \tag{3.42}$$

where λ_r and k_c are constant gains, $\hat{\mu}_x$ is an estimation of the biomass-specific growth rate $\gamma_1 r_1 + \gamma_2 r_2 + \gamma_3 r_3$ and μ_r is its desired reference value. Recall the inlet flow to the bioreactor is factorised as $F = \lambda x v$, following a closed-loop exponential feeding law, where the controller (3.42) modifies the gain λ. The estimation $\hat{\mu}_x$ can be obtained using high-gain observers from measurements of biomass (see References 7 and 9 for greater details), whereas on-line biomass measurements can be performed using several commercially available devices [71,91].

A SMRC outer loop is used to seek the unknown reference μ_r for biomass-specific growth rate. A first-order filter is employed as reference-seeking element. It is given by

$$F : \dot{\mu}_r = -a(\mu_r + w - \bar{\mu}_r) \tag{3.43}$$

where $\bar{\mu}_r$ is the desired biomass-specific growth rate when ethanol concentration does not reach its upper bound. The output of the filter, μ_r, is then used as reference by the inner-loop controller (3.42). The previous filter is driven by the discontinuous signal w taking values according to the sign of the switching function σ:

$$w = \begin{cases} \bar{\mu}_r : \sigma > 0 \\ 0 : \sigma \leq 0 \end{cases} \tag{3.44}$$

To design the switching function σ, it is illustrative to express the whole system composed of the bioreactor (3.41), the controller (3.42) and the conditioned filter (3.43) in compact state-space form. To this end, the closed-loop system state vector is defined as $\mathbf{x} \triangleq [x \; s \; \lambda \; e]^T$. The whole system description results

$$
\begin{bmatrix} \dot{x} \\ \dot{s} \\ \dot{\lambda} \\ \dot{e} \end{bmatrix} = \begin{bmatrix} \mu_x x - \lambda x^2 \\ -\mu_s x + \lambda(s_i - s)x \\ \lambda_r k_c a \mu_r \\ (\mu_e - \lambda e)x \end{bmatrix} + \begin{bmatrix} 0 \\ 0 \\ \lambda_r k_c a \\ 0 \end{bmatrix} w + \begin{bmatrix} 0 \\ 0 \\ \lambda_r k_c (\dot{\hat{\mu}}_x - a\bar{\mu}_r) \\ 0 \end{bmatrix}
$$

$$ y_1 = \mu_x $$
$$ y_2 = e $$

(3.45)

Note that from the ethanol concentration e to the discontinuous input signal w, the relative degree is $\rho = 2$ (this includes the filter, thus the constrained subsystem S_c of Chapter 2 would have unitary relative degree). Moreover, there is strong invariance with respect to the derivative of the biomass-specific growth rate $\dot{\mu}_x$, as this term, seen as a perturbation, is collinear with the control signal w.

Hence, assuming the ethanol concentration can be measured, the switching function is defined by

$$
\sigma(\bar{e}, \mathbf{x}) = e - \bar{e} + \tau \dot{e} \quad \rightarrow \quad \begin{cases} \hat{\sigma} = e - \bar{e} + \tau \hat{\dot{e}} \\ \hat{\dot{e}} = (\hat{\mu}_e - \lambda e)x \end{cases}
$$

(3.46)

where $e \le \bar{e}$ is the unavoidable constraint, and the estimation of the ethanol concentration derivative has to be added so as to achieve the transversality condition.

As the kinetic term μ_e is assumed to be unknown, the derivative $\dot{e}$ must be estimated. This can be done using different approaches. Exact differentiation can be used from measurements of e (see Reference 78). On the other hand, μ_e can be estimated using a high-gain observer in the same way as $\hat{\mu}_x$. The other terms in (3.46) are known.

The reasoning behind the control law defined by (3.44), (3.46) and the filter (3.43) is that when $\sigma > 0$, which means that either $e > \bar{e}$ or $\dot{e} > 0$ and large, μ_r decreases as $\dot{\mu}_r = -a\mu_r$. Otherwise, μ_r increases as $\dot{\mu}_r = a(\bar{\mu}_r - \mu_r)$.

Recall that SMRC is only established if the constraint is to be violated. In such a case, it can be easily deduced from (3.46) that the sliding dynamics is given by

$$
\dot{e} = \frac{1}{\tau}(\bar{e} - e)
$$

(3.47)

Thus, the ethanol concentration is smoothly and robustly driven to the constraint $\bar{e}$ with dynamics imposed by the user-defined coefficient τ. The sliding regime is abandoned as soon as the system trajectories point inside the allowed region.

Concerning the internal dynamics – recall (2.78) – it is important to remark that during sliding mode, as $\xi_s = [e \; \dot{e}]^T$ tends to $\xi_{s0} = [\bar{e} \; 0]^T$ with the fast dynamics imposed by design parameter τ, the controller gain λ in (3.41) tends to $\mu_e(\bar{e}, s, o)/\bar{e}$.

Therefore, if biomass x and substrate s are chosen as the two last coordinates in (2.73), the internal dynamics $\dot{\eta}_s = q(\xi_{s0}, \eta_s)$ is given by

$$
\eta_s : \begin{cases} \dot{x} = \mu_x x - \dfrac{\mu_e(\bar{e}, s, o)}{\bar{e}} x^2 \\[2mm] \dot{s} = -\mu_s x + \dfrac{\mu_e(\bar{e}, s, o)}{\bar{e}} (s_i - s)x \end{cases} \tag{3.48}
$$

From (3.39) and (3.40), $\mu_e(\bar{e}, s, o) = \gamma_4 r_2 - r_3 < 0$ only if $r_o > r_s \gamma_5$, but this is only possible when oxygen is in excess, in which case the SMRC loop is not active. Thus, the internal dynamics is always stable while ethanol is being bounded.

3.4.4   Simulations

The model (3.38) with parameters chosen from Table 3.1 is used to show the performance of the reference-seeking algorithm for overflow avoidance. Ethanol is assumed to be measured. Concerning this, it is important to stress that the goal is not to regulate ethanol, as in other works in the literature, but to avoid that it overcomes a given threshold. One would ideally like ethanol not to be produced, so that the cells are always under oxidative metabolism. This means that the presence of ethanol must simply be detected. This is important in practice, as cheaper ethanol probes could be used compared with the ones required for precise regulation.

Figures 3.22–3.25 show some of the results obtained when a sudden perturbation consisting of a temporal limitation in oxygen – within an otherwise situation of oxygen excess – is considered. The goal is to show in a simple way the robustness of the SMRC scheme with respect to unmeasured disturbances. In the simulation, the inner-loop control uses biomass and volume as the only measured variables, although this is not a requirement of the conditioning loop. Any controller driving the specific growth rate to the set point specified by the outer SMRC loop would serve as well. Both the inner-loop control with and without SMRC were run. The sliding function σ was determined by the choices $\tau = 1/3$ and $\bar{e} = 0.1$, considering an ethanol probe with a sensitivity of that order (e.g. in Reference 137 a probe with a sensitivity of 0.07 g/l is used). Note that since ethanol is only detected, the approach allows to consider very low thresholds for ethanol, in the range of the sensitivity of the probe.

As can be observed in Figure 3.22, the conditioning loop achieves the goal: it effectively delimits ethanol concentration, which also avoids undesired transient on biomass dynamics. In the top boxes shown in Figure 3.23, the corresponding biomass and ethanol growth rate estimations are depicted. Note that when no reference conditioning is used, activation of the fermentative path leads to a temporal decrease in biomass growth rate, followed by a growth rate overshoot as ethanol in excess is oxidised (dashed line). If SMRC is added, no ethanol in excess is produced. Thus, the specific biomass growth rate is the maximum compatible with the constraint on ethanol (solid line). The bottom half in Figure 3.23 presents the evolution of the switching function σ and the control action λ. As observed in the right subplot, the sliding regime is reached when oxygen limitation activates the fermentative path. As soon as the process conditions allow it, the sliding regime is left. In Figure 3.24, the

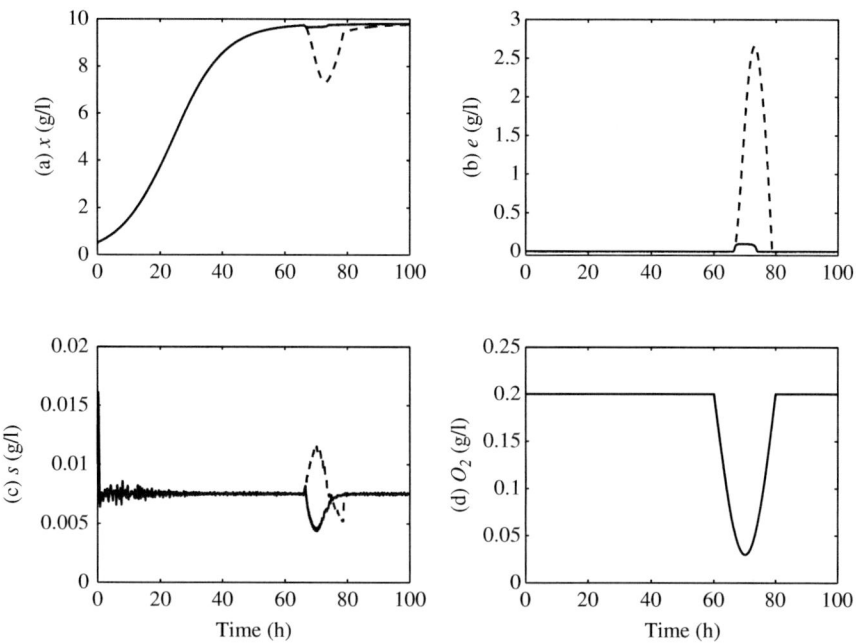

Figure 3.22   Evolution of the biomass, ethanol, glucose and oxygen concentrations for the system with (solid) and without (dashed) SMRC

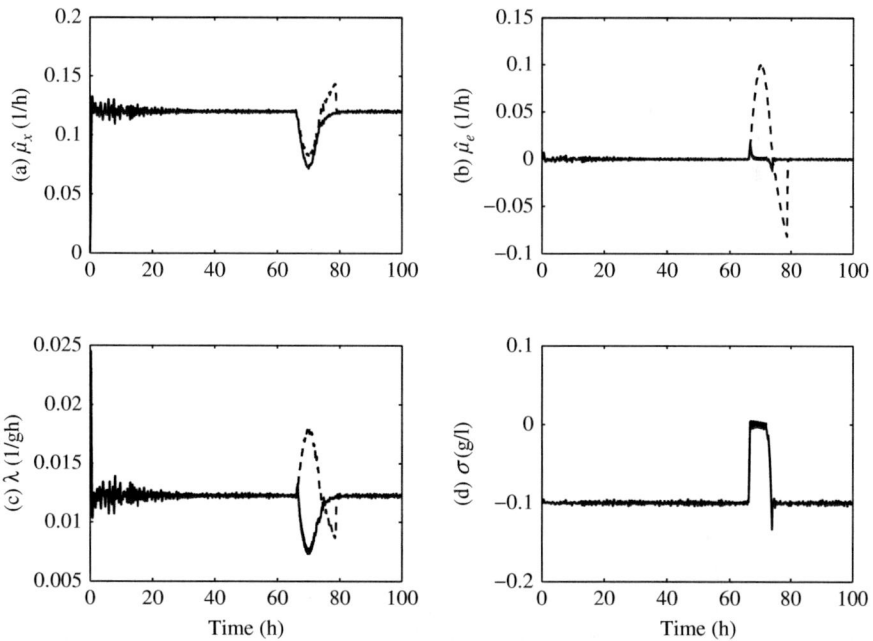

Figure 3.23   Biomass/ethanol rates and controller gain λ with (solid) and without (dashed) SMRC, and sliding function σ

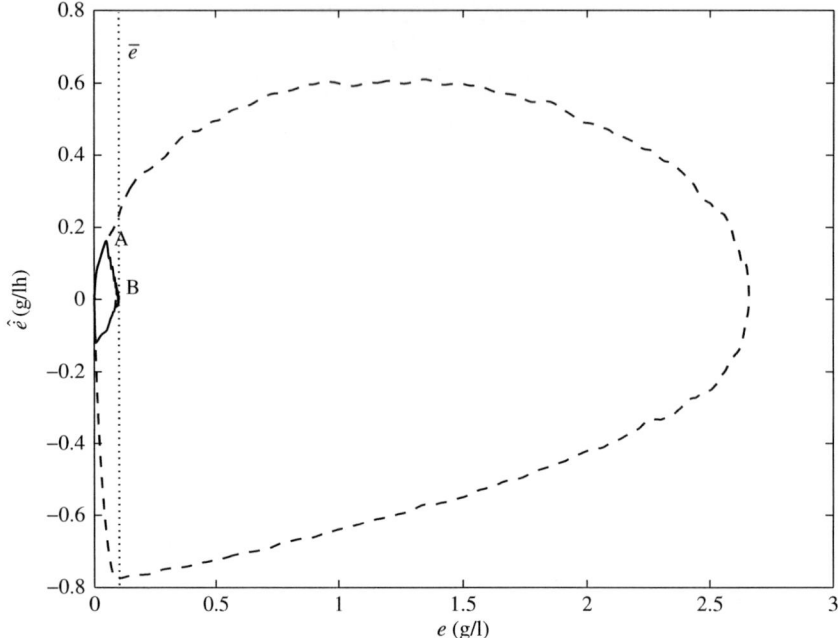

*Figure 3.24   Ethanol trajectories on phase plane with (solid) and without (dashed)
sliding mode reference conditioning*

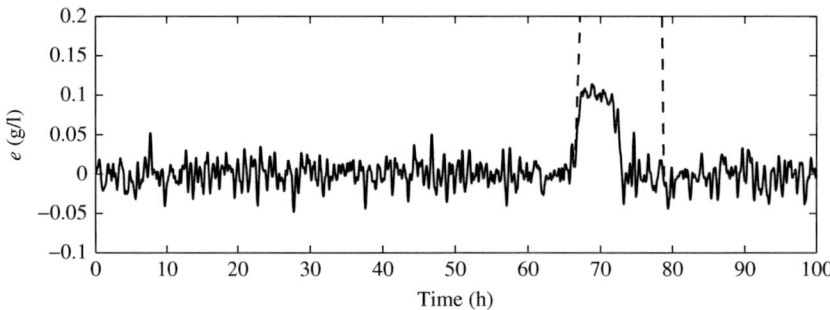

*Figure 3.25   Zoomed ethanol concentration with noise component of the order of
ethanol probes sensitivity, with (solid) and without (dashed) the
SMRC scheme*

ethanol and its (estimated) derivative trajectories highlight the constraint satisfaction
($e \leq 0.1$) achieved with the SMRC approach (solid line). Sliding regime is established in A, and thereafter ethanol concentration tends to its limit value with the
dynamics imposed by τ (from A to B). Finally, Figure 3.25 presents a zoomed plot of
ethanol concentration when a noise component of the order of referenced ethanol
probes (0.07 g/l) is considered. It is important to observe that, although the noise
effectively affects the reaching dynamics to the constraint, the method manages to
maintain ethanol concentration below the desired threshold.

Chapter 4
Relevant tools for dynamic decoupling

From this chapter, we will deal with multivariable dynamical systems, already introduced in Chapter 1. We are particularly interested in improving the decoupling of multiple-input multiple-output (MIMO) systems in the presence of different types of constraints, that is, to reduce interactions among the loops as much as possible when the plant under control is subjected to either physical, structural or dynamic constraints.

To this end, some important tools for control and dynamic decoupling of ideally unconstrained multivariable systems are presented in this chapter. First, we recall some basic concepts of MIMO systems analysis. Then, the affine parameterisation of all the controllers which internally stabilise the feedback system is introduced, together with the closely related Internal Model Control (IMC). Based on this control strategy, some design procedures are described in order to achieve the dynamic decoupling of stable minimum phase (MP) systems, non-minimum phase (NMP) systems and unstable systems. Finally, the effects and limitations of diagonal (full) dynamic decoupling in systems with right-half plane (RHP) zeros are briefly discussed.

4.1 Preliminary concepts

4.1.1 Multivariable system models

We start by reviewing two different ways of describing a MIMO system.

4.1.1.1 State-space models

MIMO linear systems as well as single-input single-output (SISO) systems can be represented in the time domain by means of ordinary differential equations in the state space. The only difference is the dimension of the input, the output and the state-space matrices. In the case of a linear MIMO system with input $u(t) \in \mathbb{R}^m$ and output $y(t) \in \mathbb{R}^m$, the state-space model can be written as

$$\dot{x}(t) = Ax(t) + Bu(t), \quad x(t_0) = x_0 \tag{4.1}$$

$$y(t) = Cx(t) + Du(t) \tag{4.2}$$

where $x(t) \in \mathbb{R}^n$ is the state vector, $A \in \mathbb{R}^{n \times n}$, $B \in \mathbb{R}^{n \times m}$, $C \in \mathbb{R}^{m \times n}$ and $D \in \mathbb{R}^{m \times m}$ are constant matrices and $x_0 \in \mathbb{R}^n$ is the state vector at instant $t = t_0$.

4.1.1.2   Transfer matrices

Computing the Laplace transform of (4.1) and (4.2) and taking $x(0) = 0$ yields

$$y(s) = \left[C(sI - A)^{-1}B + D \right] u(s) \tag{4.3}$$

The system transfer matrix is then defined as

$$P(s) \triangleq C(sI - A)^{-1}B + D \tag{4.4}$$

Denoting with $p_{ij}(s)$ the transfer function between the jth component of $u(s)$ and the ith component of $y(s)$, $P(s)$ can be written in terms of the individual transfer functions as

$$P(s) = \begin{bmatrix} p_{11}(s) & p_{12}(s) & \cdots & p_{1j}(s) & \cdots & p_{1m}(s) \\ p_{21}(s) & p_{22}(s) & \cdots & p_{2j}(s) & \cdots & p_{2m}(s) \\ \vdots & \vdots & \cdots & \vdots & \cdots & \vdots \\ p_{i1}(s) & p_{i2}(s) & \cdots & p_{ij}(s) & \cdots & p_{im}(s) \\ \vdots & \vdots & \cdots & \vdots & \cdots & \vdots \\ p_{m1}(s) & p_{m2}(s) & \cdots & p_{mj}(s) & \cdots & p_{mm}(s) \end{bmatrix} \tag{4.5}$$

Note that $P(s) \in \mathbb{R}^{m \times m}(s)$, the set of $m \times m$ matrices with elements in $\mathbb{R}(s)$, $\mathbb{R}(s)$ being the field of rational transfer functions in s with real coefficients. $P(s)$ will be assumed nonsingular except for isolated values of s.

We recall now the definition of proper, strictly proper and improper transfer matrices:

Definition 4.1 (Proper, strictly proper and improper transfer matrices): *$P(s) \in \mathbb{R}^{m \times m}(s)$ is said to be proper if all its elements are proper transfer functions. In particular, a proper transfer matrix $P(s)$ is said to be strictly proper if $\lim_{s \to \infty} \det(P(s)) = 0$. All transfer matrices which are not proper are improper.*

4.1.2   Multivariable poles and zeros

In a multivariable system, the presence and position of the zeros cannot always be determined from simply taking a glance at the elements of the transfer matrix $P(s)$. In fact, the zeros of the individual transfer functions $p_{ij}(s)$ do not cancel in general the output vector $y(t)$ for non-identically zero input $u(t)$. That is, they do not have on the multivariable system the blocking effect, which is characteristic of the zeros in a SISO system. A definition that preserves this feature for the MIMO system zeros, because of which they are usually called *transmission zeros*, is the one that follows [83].

Definition 4.2 (Multivariable zeros): *z is a zero of the transfer matrix $P(s)$ if the rank of $P(z)$ is lower than the normal rank of $P(s)$, where the normal rank of $P(s)$ is the rank of $P(s)$ for every value of s except for a finite number of singularities.*

From this definition, there will exist constant and non-zero vectors $v \in \mathbb{R}^m$ and $h \in \mathbb{R}^m$, known as *input* and *output direction* of the zero at $s = z$, respectively, such that

$$P(z)v = 0 \in \mathbb{R}^m \tag{4.6}$$

$$h^\mathsf{T} P(z) = 0 \in \mathbb{R}^m \tag{4.7}$$

Note that the vectors v and h are part of the null spaces generated by the columns and rows of $P(z)$, respectively. The dimension of these null spaces is known as the *geometric multiplicity* of the zero, which is equal to the number of linearly independent vectors that satisfy (4.6) or (4.7) and depends on the rank loss of $P(s)$ when evaluated at $s = z$.

Without being so rigorous, the poles of a multivariable system can be defined as those finite values of $s = p$ where $P(s)$ has a singularity (e.g. is 'infinite'). Hence, given a pole of $P(s)$ at $s = p$, it would be

$$P(p)v_p = \infty \tag{4.8}$$

$$h_p^\mathsf{T} P(p) = \infty \tag{4.9}$$

where v_p and h_p are the input and output pole directions at $s = p$, respectively.

For the case of square systems, the zeros and poles of $P(s)$ can be computed from the zeros and poles of the determinant of $P(s)$. However, because of the aforementioned directions associated with multivariable zeros and poles, one must be sure of avoiding pole/zero cancellations when forming the determinant to use this simple method. Otherwise, poles and zeros on the same (frequency) location but with different directions (spatial location) could be wrongly cancelled.[1]

Example 4.1: The determinant of the system

$$P(s) = \begin{bmatrix} \dfrac{s+2}{s+1} & 0 \\ 0 & \dfrac{s+1}{s+2} \end{bmatrix} \tag{4.10}$$

is $det(P(s)) = 1$, although the system clearly has poles at $s = -1$ and $s = -2$ and multivariable zeros at $s = -1$ and $s = -2$. This is a case in which zeros and poles in different parts of the system are cancelled when forming the determinant, and thus, it is not possible to obtain them from the condition $det(P(s)) = 0$. They are poles and zeros with the same location but with different direction. In effect, the pole at $s = -1$ (and the zero at $s = -2$) has direction $v_p = h_p = [1\ 0]^\mathsf{T}$, whereas the zero at $s = -1$ (and the pole at $s = -2$) has direction $v = h = [0\ 1]^\mathsf{T}$.

Definition 4.3 (Non-minimum phase system): *It is said that a multivariable system is non-minimum phase (NMP) if its transfer matrix $P(s)$ has zeros in the*

---

[1] The reader is referred to Reference 118 for more details on multivariable zeros computation.

right-half plane (RHP) or if there exists a common time delay that can be factored out of every matrix element $p_{ij}(s)$.

Although being a direct consequence of the previous definitions, it is important to remark here that the multivariable or transmission zero locations bear no relation to the locations of the zeros of the individual SISO transfer functions that comprise the MIMO transfer matrix. Then, it is possible for a MIMO system to be NMP although having all its individual transfer functions of minimum phase (MP), and vice versa. The following two examples illustrate both cases.

Example 4.2: Consider the system

$$P_1(s) = \frac{1}{(1+s)^2} \begin{bmatrix} s+3 & 2 \\ 3 & 1 \end{bmatrix} \tag{4.11}$$

Clearly, none of the individual transfer functions has a zero in the RHP. However, when considering the MIMO system, it can be observed that it has a finite zero at $s = +3$ with output direction $h = [1 \ -2]^T$. Therefore, (4.11) is an NMP multivariable system even though all its SISO transfer functions are MP. This fact imposes performance limitations to control the MIMO system, while each SISO subsystem can be easily controlled to achieve the desired closed-loop dynamics.

Example 4.3: Take another system now

$$P_2(s) = \frac{1}{(s+1)(s+2)} \begin{bmatrix} s-1 & s \\ -6 & s-2 \end{bmatrix} \tag{4.12}$$

We have here the opposite case with respect to the previous example. The multivariable system has no zeros in the RHP, since they are at $s = -1$ and $s = -2$. Nevertheless, $p_{11}(s)$ and $p_{22}(s)$ have RHP zeros at $s = 1$ and $s = 2$, respectively. Observe that for a plant with these characteristics, there will be greater performance limitations to control each SISO loop separately (because of the NMP zeros) than to control the complete MIMO system, whose closed loop can have greater bandwidth than the one of each individual loop.

4.1.3 Closed-loop transfer matrices

For the rest of the book, we will consider a unitary negative feedback control structure as the one shown in Figure 4.1, where $P(s)$ represents the transfer matrix of the plant to be controlled, with m inputs and m outputs.

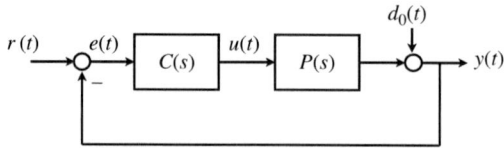

Figure 4.1 Basic structure of a MIMO control loop

The main control objective will be that the output vector $y(t) \in \mathbb{R}^m$, which contains the variables to be controlled, follows the desired trajectories included in the reference vector $r(t) \in \mathbb{R}^m$ (tracking objective). It will also be considered the case in which it is aimed that the controlled variables remain as close as possible to constant *set points* in presence of external disturbances, represented by the vector $d_0(t)$ (regulation objective).

The controlled variables can be expressed as

$$y(s) = P(s)u(s) + d_0(s) \tag{4.13}$$

and so

$$y(s) = P(s)C(s)[r(s) - y(s)] + d_0(s) \tag{4.14}$$

Thus, the closed-loop response will be given by

$$y(s) = [I + P(s)C(s)]^{-1}P(s)C(s)r(s) + [I + P(s)C(s)]^{-1}d_0(s) \tag{4.15}$$

The resultant closed-loop transfer matrices

$$S(s) \overset{\Delta}{=} [I + P(s)C(s)]^{-1} \tag{4.16}$$

$$T(s) \overset{\Delta}{=} [I + P(s)C(s)]^{-1}P(s)C(s)$$
$$= P(s)C(s)[I + P(s)C(s)]^{-1} \tag{4.17}$$

are known as *sensitivity function matrix* $S(s)$ and *complementary sensitivity function matrix* $T(s)$. Note that the term *sensitivity* comes from the fact that $((\partial T/T)/(\partial P/P)) = S(s)$, whereas the term *complementary* comes from

$$S(s) + T(s) = [I + P(s)C(s)]^{-1}P(s)C(s) + [I + P(s)C(s)]^{-1}$$
$$= [I + P(s)C(s)]^{-1}[I + P(s)C(s)] = I \tag{4.18}$$

Hence, the control error $e(s)$ is

$$e(s) = r(s) - y(s) = r(s) - T(s)r(s) - S(s)d_0(s)$$
$$= [I - T(s)]r(s) - S(s)d_0(s) = S(s)r(s) - S(s)d_0(s) \tag{4.19}$$

and the controller output is given by

$$u(s) = C(s)S(s)r(s) - C(s)S(s)d_0(s) \tag{4.20}$$

The transfer matrix from $r(s)$ to $u(s)$ is usually called *input sensitivity matrix* and denoted by

$$S_u(s) = C(s)[I + P(s)C(s)]^{-1} \tag{4.21}$$

4.1.4   Internal stability

It is frequently understood that a system is stable provided its transfer function or matrix from a particular input (reference) to a particular output (controlled variable) has all its poles in the left-half plane (LHP). However, the interconnection of systems as shown in the control loop shown in Figure 4.1 may give rise to internal unstable modes, hidden by potential RHP pole/zero cancellations. Then, it is useful to establish when a control system is internally stable and not only bounded-input bounded-output (BIBO) stable, that is, to require stability for signals injected or measured at any point of the system.

Definition 4.4 (Internal Stability): *A closed-loop control system is said to be internally stable if and only if all the signals in the loop are bounded for every set of bounded inputs.*

Then, for the control loop in Figure 4.1 to be internally stable, $e(t)$, $u(t)$ and $y(t)$ must be bounded for every bounded input $r(t)$ and disturbance $d_0(t)$.

Lemma 4.1: *The (nominal) feedback system given in Figure 4.1 is internally stable if and only if*

$$S(s) = [I + P(s)C(s)]^{-1} \tag{4.22}$$

$$T(s) = [I + P(s)C(s)]^{-1}P(s)C(s) \tag{4.23}$$

$$S_u(s) = C(s)[I + P(s)C(s)]^{-1} \tag{4.24}$$

are all stable.

Note that, obviously, stability of $S(s)$ implies stability of $T(s)$, and vice versa.

Remark 4.1: *Observe that in the loop given in Figure 4.1, no input disturbances have been considered. If this were done, the transfer matrices $S_i(s) = [I + P(s)C(s)]^{-1}P(s) = P(s)[I + C(s)P(s)]^{-1} = S(s)P(s)$ and $S'(s) = [I + C(s)P(s)]^{-1}$ should also be stable for the loop to be internally stable.*

4.2   MIMO controller parameterisation and approximate inverses

A basic concept that is present in the great majority of the problems addressed by control theory is the fact that the control of a system depends, explicitly or implicitly, on the inversion of the plant model. In the following text, this point is further discussed, together with an alternative methodology for the design of feedback controllers.

4.2.1   Stabilising controller parameterisation

It seems natural to think that the control vector $u(s)$, which is needed to achieve a given response on the controlled variable $y(s)$, could be produced in an open-loop configuration from the reference vector $r(s)$, as shown in Figure 4.2.

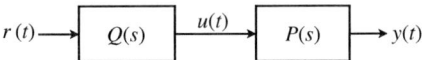

Figure 4.2   Open-loop control scheme

Calling $Q(s)$ the precompensator transfer matrix, which connects $r(s)$ to $u(s)$, this scheme leads to an input–output transfer matrix affine in $Q(s)$:

$$T(s) = P(s)Q(s) \tag{4.25}$$

Equation (4.25) reveals the importance of inversion. Indeed, if $Q(s)$ were able to invert the plant model, the so-called perfect control would be reached, i.e. $T(s) = I$. Although unfortunately this is not possible in practice, we will study further below how to compute approximate inverses of the plant models.

In the previous section, we have seen that the closed-loop transfer matrix of a conventional unitary feedback configuration, as the one given in Figure 4.1, can be expressed in terms of the controller $C(s)$ by

$$T(s) = P(s)C(s)[I + P(s)C(s)]^{-1} = [I + P(s)C(s)]^{-1}P(s)C(s) \tag{4.26}$$

This expression depends on a non-linear manner of $C(s)$, which hinders the controller tuning to attain the desired closed-loop specifications. From (4.25) and (4.26), we have

$$Q(s) = C(s)[I + P(s)C(s)]^{-1} \equiv S_u(s) \tag{4.27}$$

Then, the following parameterisation of the controller $C(s)$ in terms of the affine parameter $Q(s)$ can be derived

$$C(s) = Q(s)[I - Q(s)P(s)]^{-1} = [I - Q(s)P(s)]^{-1}Q(s) \tag{4.28}$$

which is known as Q-parameterisation or *Youla parameterisation* [154,157]. This result is formalised in the following lemma.

Lemma 4.2 (Affine parameterisation of stable systems): *Consider a plant with a stable nominal model P(s) controlled by a negative unitary feedback controller C(s), as shown in Figure 4.1. Then, the nominal feedback system is internally stable if and only if C(s) can be parameterised as in (4.28), with Q(s) being any stable and proper transfer matrix.*

Proof:

From (4.27), the sensitivity-transfer matrices (4.22)–(4.24) of Lemma 4.1 (which must be stable for the loop to be internally stable) result:

$$C(s)[I + P(s)C(s)]^{-1} = Q(s) \tag{4.29}$$

$$[I + P(s)C(s)]^{-1} = I - P(s)Q(s) \tag{4.30}$$

$$P(s)C(s)[I + P(s)C(s)]^{-1} = P(s)Q(s) \tag{4.31}$$

Given that $P(s)$ is stable, the latter are all stable and proper transfer matrices if and only if $Q(s)$ is stable and proper.

Remark 4.2: *For systems with disturbances at the plant input, the Youla parameterisation does also guarantee internal stability of the closed loop, since $P(s)[I + C(s)P(s)]^{-1} = P(s)[I - Q(s)P(s)]$ and $[I + C(s)P(s)]^{-1} = I - Q(s)P(s)$ are stable and proper if $P(s)$ and $Q(s)$ are stable and proper as well.*

In the following text, we will try to take advantage of the simplicity of (4.25) and the idea of looking for an approximate inverse of the plant to design $Q(s)$ first, and then – if necessary – to obtain $C(s)$ by means of (4.28).

4.2.2   Internal model control

From (4.25), one could think of simplifying the design procedure to obtain a given input–output closed-loop dynamics in $T(s)$ by exploiting the parameter $Q(s)$. However, in the open-loop structure given in Figure 4.2, there is no information about the controlled variables; thus, the effects of potential disturbances and/or uncertainties are completely unknown and there are no tools to take them into account.

A closed-loop control strategy that takes advantage of the design simplicity of the parameter $Q(s)$ is the so-called IMC [88]. Figure 4.3 presents a block diagram of this strategy. The area enclosed with dotted line represents the controller to be implemented. As can be seen, the name of this control strategy comes from the fact that the complete controller includes not only $Q(s)$ but also the plant model $P(s)$ [33–35].

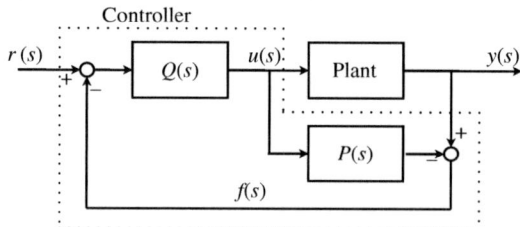

Figure 4.3   Internal Model Control (IMC) structure

Apart from the design facilities provided by the IMC control, another important advantage of this control methodology is that, given an open-loop stable plant model $P(s)$, the internal stability of the feedback system is ensured by simply choosing a stable $Q(s)$, as verified by (4.29)–(4.31).

The IMC configuration can be easily obtained from the basic unitary feedback configuration given in Figure 4.1. Figure 4.4 shows the equivalence between both configurations: adding and subtracting in the conventional feedback scheme the signal $P(s)u(s)$ and using the expression of $Q(s)$ obtained in (4.27) yields the IMC structure.

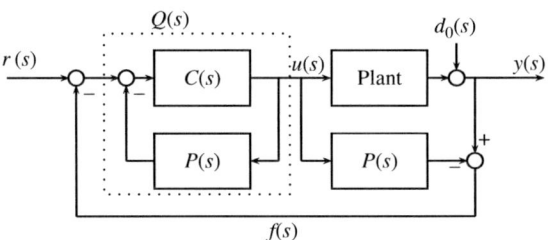

Figure 4.4   Equivalence between IMC control and classic feedback configuration

It is worth highlighting that in this configuration the feedback signal is given by

$$f(s) = [P(s) - \tilde{P}(s)]u(s) + d_0(s) \tag{4.32}$$

where the real plant to be controlled has been denoted with $\tilde{P}(s)$. That is, the feedback signal $f(s)$ does only depend on the modelling errors and external disturbances.[2] If the model were exact, i.e. $P(s) = \tilde{P}(s)$, and there were no disturbances $d_0(s)$, the feedback signal $f(s)$ would be zero and the control system would work in an open-loop fashion.

The previous reasoning illustrates in a simple manner why feedback is necessary in a control system: because of model uncertainty and output disturbances.

4.2.3   Interactor matrices

We have already seen how to use $Q(s)$ as the system controller in a closed-loop configuration. We now need to find the way of approximating a proper inverse of the plant model. With this aim, we study the interactor matrices.

In linear SISO systems, it is said that the relative degree (ρ) of a transfer function $P_1(s) \in \mathbb{R}(s)$ is the degree of a polynomial $pol(s)$ such that

$$\lim_{s \to \infty} pol(s)P_1(s) = K, \quad \text{where} \quad 0 < |K| < \infty \tag{4.33}$$

---

[2] Note that $f(s)$ is the perturbed component of the output $y(s)$, and then it can be seen as the disturbance signal d affecting SMRC schemes in Figures 2.1 and 2.9 (see also Figure 2.3).

The polynomial $pol(s)$ is such that $pol(s)P_1(s)$ is biproper, i.e. $[pol(s)P_1(s)]^{-1}$ is also proper. This polynomial is unique if it is required to belong to the class of polynomials $\wp = \{s^k | k \in \mathbb{N}\}$.

For MIMO systems, each entry of the transfer matrix $P(s) \in \mathbb{R}^{m \times m}(s)$ may have a different relative degree. Hence, to generate a condition analogous to (4.33), it will be necessary to define a matrix that takes into account the individual transfer functions $p_{ij}(s)$ and their different relative degrees.

In effect, given $P(s) \in \mathbb{R}^{m \times m}(s)$, there will exist matrices $\xi_l(s)$ and $\xi_r(s)$ such that the following conditions are met:

$$\lim_{s \to \infty} \xi_l(s)P(s) = K_l, \quad \text{where} \quad 0 < |det(K_l)| < \infty \tag{4.34}$$

$$\lim_{s \to \infty} P(s)\xi_r(s) = K_r, \quad \text{where} \quad 0 < |det(K_r)| < \infty \tag{4.35}$$

It is important to remark here that although (4.34) and (4.35) state that both $\xi_l(s)P(s)$ and $P(s)\xi_r(s)$ are biproper, in MIMO system this does not imply that each element of the transfer matrix must be a biproper transfer function.

The previous result is formalised in the following theorem:

Theorem 4.1: *Consider a square transfer matrix $P(s) \in \mathbb{R}^{m \times m}$, non-singular almost everywhere in s. Then, there exist unique transfer matrices $\xi_l(s)$ and $\xi_r(s)$ such that (4.34) and (4.35) are verified and*

$$\xi_l(s) = H_l(s)D_l(s), \quad \text{with} \tag{4.36}$$

$$D_l(s) = diag(s^{p_1}, \ldots, s^{p_m}) \tag{4.37}$$

$$H_l(s) = \begin{bmatrix} 1 & 0 & \cdots & \cdots & 0 \\ h_{21}^l(s) & 1 & \cdots & \cdots & 0 \\ h_{31}^l(s) & h_{32}^l(s) & \ddots & & \vdots \\ \vdots & \vdots & & \ddots & \vdots \\ h_{m1}^l(s) & h_{m2}^l(s) & \cdots & \cdots & 1 \end{bmatrix} \tag{4.38}$$

$$\xi_r(s) = D_r(s)H_r(s), \quad \text{with} \tag{4.39}$$

$$D_r(s) = diag(s^{q_1}, \ldots, s^{q_m}) \tag{4.40}$$

$$H_r(s) = \begin{bmatrix} 1 & h_{12}^r(s) & h_{13}^r(s) & \cdots & h_{1m}^r(s) \\ 0 & 1 & h_{23}^r(s) & \cdots & h_{2m}^r(s) \\ \vdots & \vdots & \ddots & & \vdots \\ \vdots & \vdots & & \ddots & \vdots \\ 0 & 0 & \cdots & \cdots & 1 \end{bmatrix} \tag{4.41}$$

with $h_{ij}^l(s)$ and $h_{ij}^r(s)$ polynomials in s that satisfy $h_{ij}^l(0) = 0$ and $h_{ij}^r(0) = 0$.

The transfer matrices $\xi_l(s)$ and $\xi_r(s)$ are known as left and right interactor matrices, respectively.

The relevance of the interactor matrices or just *interactors* relies on the fact that they define the relative degree (or infinite zeros) structure of a multivariable system. For a feedback control system, $\xi_l(s)$ and $\xi_r(s)$ define the minimum relative degree structure of any complementary sensitivity achievable with a proper controller.

Remark 4.3: *Note that although in (4.37) and (4.40) the diagonal elements of the matrices $D_l(s)$ and $D_r(s)$ belong to the class of polynomials $\wp = \{s^k | k \in \mathbb{N}\}$, they can be chosen as arbitrary polynomials of degree $p_1, \ldots, p_m$ and $q_1, \ldots, q_m$, respectively. In particular, they can always be chosen as Hurwitz polynomials, which is of great importance since in this way the inverses of $\xi_l(s)$ and $\xi_r(s)$ will be stable transfer matrices.*

4.2.3.1 Constructing interactor matrices

We now describe a procedure to construct the left interactor $\xi_l(s)$ that can be easily implemented in a computer algorithm. The right interactor $\xi_r(s)$ can be obtained following the same procedure but interchanging the rows and columns indexes. For greater details on how to compute interactors and their properties, see References 49, 105 and 150.

Consider the *i*th row of $P(s)$, to be denoted as $[P(s)]_{i*}$. Then, there exists a minimum non-negative integer n_i, such that

$$\lim_{s \to \infty} s^{n_i} [P(s)]_{i*} = f_i^T \tag{4.42}$$

with f_i^T being a constant row vector, non-zero and with finite entries.

To construct $\xi_l(s)$, one should proceed in the following manner:
Step 1: Choose the first row of $\xi_l(s)$, $[\xi_l(s)]_{1*}$, as

$$[\xi_l(s)]_{1*} = [s^{n_1} \ 0 \ 0 \ldots 0] \tag{4.43}$$

Thus, $[K_l]_{1*} = \lim_{s \to \infty} [\xi_l(s)]_{1*} P(s) = f_1^T$.
Step 2: Consider the second row vector in (4.42), f_2^T. If f_2^T is linearly independent of f_1^T, choose the second row of $\xi_l(s)$, $[\xi_l(s)]_{2*}$, as

$$[\xi_l(s)]_{2*} = [0 \ s^{n_2} \ 0 \ldots 0] \tag{4.44}$$

which leads to $[K_l]_{2*} = \lim_{s \to \infty} [\xi_l(s)]_{2*} P(s) = f_2^T$.
Step 3: If f_2^T is linearly dependent on f_1^T, there will exist a constant β_2^1 such that $f_2^T = \beta_2^1 f_1^T$. If $[\xi_l(s)]_{2*}$ were chosen as in (4.44), K_l in (4.34) would be singular. To avoid this, the following row vector is formed

$$[\xi_l(s)]_{2*}^1 = s^{n_2^1} \left([0 \ s^{n_2} \ 0 \ldots 0] - \beta_2^1 [\xi_l(s)]_{1*}\right) \tag{4.45}$$

with n_2^1 being the unique integer such that $\lim_{s \to \infty} [\xi_l(s)]_{2*}^1 P(s) = f_2^{1T}$, a non-zero and finite row vector.

If f_2^{1T} results linearly independent of f_1^T, one takes $[\xi_l(s)]_{2*} = [\xi_l(s)]_{2*}^1$. Conversely, if $f_2^{1T} = \beta_2^2 f_1^T$, with β_2^2 constant, then the process given in (4.45) is repeated, but now between $[\xi_l(s)]_{2*}^1$ and $[\xi_l(s)]_{1*}$:

$$[\xi_l(s)]_{2*}^2 = s^{n_2^2}\left([\xi_l(s)]_{2*}^1 - \beta_2^2[\xi_l(s)]_{1*}\right) \tag{4.46}$$

This procedure should be repeated until linear independence is reached. Note that this will always happen when dealing with models of real systems, since they are represented by strictly proper transfer functions.

Step 4: Obtain the subsequent rows of $\xi_l(s)$ analogously.

Example 4.4: Consider the transfer matrix

$$P(s) = \frac{1}{(s+5)^3}\begin{bmatrix} (s+5)^2 & (s+5) \\ 3(s+5) & 1 \end{bmatrix} \tag{4.47}$$

To construct the left interactor $\xi_l(s)$, we follow the above-described steps:
Step 1: As in this case $n_1 = 1$,

$$[\xi_l(s)]_{1*} = [s^{n_1}\ 0\ 0\ldots 0] = [s\ 0] \tag{4.48}$$

Then,

$$[K_l]_{1*} = \lim_{s\to\infty}\ [\xi_l(s)]_{1*}P(s) = f_1^T = [1\ 0] \tag{4.49}$$

Step 2: From the second row of $P(s)$, we have that $n_2 = 2$ and $f_2^T = [3\ 0]$.
Step 3: Since f_2^T is linearly dependent on f_1^T (with $\beta_2^1 = 3$), we must choose the second row of $\xi_l(s)$ as

$$[\xi_l(s)]_{2*}^1 = s^{n_2^1}\left([0\ s^{n_2}] - \beta_2^1[\xi_l(s)]_{1*}\right) = [-3s^{(n_2^1+1)}\ s^{(n_2^1+2)}] \tag{4.50}$$

where n_2^1 is found to make

$$f_2^{1T} = \lim_{s\to\infty}\ [\xi_l(s)]_{2*}^1 P(s) = \lim_{s\to\infty}\left[\frac{-15s^{n_2^1+1}}{(s+5)^2}\quad \frac{-2s^{n_2^1+2}-15s^{n_2^1+1}}{(s+5)^3}\right] \tag{4.51}$$

a non-zero and finite row vector. Hence, we obtain $n_2^1 = 1$ and $f_2^{1T} = [-15 - 2]$, which is linearly independent of f_1^T.

Finally, the left interactor results:

$$\xi_l(s) = \begin{bmatrix} s & 0 \\ -3s^2 & s^3 \end{bmatrix} \tag{4.52}$$

Note that, although the above example has required some calculation to illustrate the procedure, in a great number of plant models the interactors can be determined by simple inspection of their relative degree, as we will verify in other examples further below.

Remark 4.4: *We have seen that the interactors describe the structure of the zeros at ∞ of a multivariable system. Then, the same basic idea as the one of Theorem 4.1 can be used to describe the structure of finite zeros in the RHP of a given plant. The resultant matrices are known as z-interactors [113,147].*

4.2.4 Approximate model inverses

We describe now how to obtain an approximate proper inverse of a strictly proper plant model $P(s)$ by means of the interactor matrices.

To this end, it is useful to define $\Lambda_l(s) \stackrel{\Delta}{=} \xi_l(s)P(s)$ and $\Lambda_r(s) \stackrel{\Delta}{=} P(s)\xi_r(s)$. Note that both $\Lambda_l(s)$ and $\Lambda_r(s)$ are biproper transfer matrices with non-singular gain at high frequencies. Then, they can be employed to construct approximate inverses of $P(s)$.

A right approximate inverse can be taken as

$$P_R(s) = \Lambda_l^{-1}(s)\xi_l(0) = [\xi_l(s)P(s)]^{-1}\xi_l(0) \tag{4.53}$$

since it satisfies

$$P(s)P_R(s) = [\xi_l(s)]^{-1}\xi_l(0) \tag{4.54}$$

which is lower triangular and equal to the identity matrix at $s = 0$.

Similarly,

$$P_L(s) = \xi_r(0)\Lambda_r^{-1}(s) = \xi_r(0)[P(s)\xi_r(s)]^{-1} \tag{4.55}$$

is a proper left approximate inverse, since

$$P_L(s)P(s) = \xi_r(0)[\xi_r(s)]^{-1} \tag{4.56}$$

is upper triangular and equal to the identity at $s = 0$.

4.3 Dynamic decoupling of MIMO systems

We will next study some design procedures in order to attain the dynamic decoupling of stable minimum phase, stable non-minimum phase and/or unstable systems.

The closed-loop decoupling is an often design requirement in MIMO control. There are different degrees of decoupling, ranging from partial to full decoupling, and from static (only in steady state) to dynamic decoupling (at all frequencies). Clearly, full dynamic decoupling is the strongest demand. It implies that any change in the set-point value of a controlled variable of the system leads to a response only in that process variable, whereas all the other controlled variables remain unaffected. Ideally, it converts a MIMO system into a set of SISO systems, and is therefore an approximation to the so-called perfect control, in which $T(s) = I$.

The advantages of full dynamic decoupling are intuitive. For instance, one may want to change the temperature of a process keeping the pressure constant or to modify the pitch angle of a plane without affecting its flight height. Indeed, some degree of decoupling (over a given bandwidth) is always required, at least around $s = 0$. Thus, the decoupling techniques to be studied are applicable to almost all MIMO control problems.

Although several methods to dynamically decouple linear multivariable systems have been reported in the literature [22,28,57,82,94,144,145,153], we will mainly follow the analysis and design presented in References 48 and 49, since apart from being one of the most widely spread methodologies, it takes advantage of the controller parameterisation studied in Section 4.2, which will be very useful for the subsequent chapters.

4.3.1   Minimum phase systems

For the nominal dynamic decoupling of stable systems, we will make use of the affine parameterisation, IMC control and interactor matrices presented in the previous section. When considering stable systems with all their zeros in the LHP, the complete decoupling can be achieved by choosing an IMC controller $Q(s)$ as follows:

$$Q(s) = P^{-1}(s)T_D(s) \qquad (4.57)$$

where

$$T_D(s) = diag(t_1(s), t_2(s), \ldots, t_m(s)) \qquad (4.58)$$

and $t_1(s), t_2(s), \ldots, t_m(s)$ are stable and proper transfer functions with unitary steady-state gain. As is obvious, the controller $Q(s)$ chosen in (4.57) leads to the complementary sensitivity matrix

$$T(s) = P(s)Q(s) \qquad (4.59)$$

$$= diag(t_1(s), t_2(s), \ldots, t_m(s)) \qquad (4.60)$$

which is diagonal as we aimed.

For this control to be realisable – and thus implementable – it must be guaranteed that the controller $Q(s)$ is proper. Note that according to (4.28), if the IMC controller $Q(s)$ is proper, then the controller $C(s)$ of the conventional unitary feedback configuration will also be proper. In general, biproper controllers are aimed to avoid unnecessary delays at high frequencies in the control system. To this end, the transfer functions $t_i(s)$ must have relative degree equal to the degree of the corresponding column $[\xi_l(s)]_{*i}$ of the left interactor of $P(s)$.

The main limitation of the design proposed in (4.57) is that the resultant controller $Q(s)$ is stable, provided $P(s)$ is MP.

Example 4.5: Consider a MIMO process with the following model

$$P(s) = \frac{1}{4s^2 + 4s + 1} \begin{bmatrix} -s+2 & 2s+1 \\ -3 & -s+2 \end{bmatrix} \tag{4.61}$$

which is stable and MP (see Section 4.1.2).

Making use of the procedure given in Section 4.2.3, we have

$$\xi_l(s) = \xi_r(s) = \begin{bmatrix} s & 0 \\ 0 & s \end{bmatrix} \tag{4.62}$$

Assume now that the following closed-loop complementary sensitivity is aimed

$$T(s) = T_D(s) = \begin{bmatrix} \dfrac{2s+4}{s^2+3s+4} & 0 \\ 0 & \dfrac{2s+4}{s^2+3s+4} \end{bmatrix} \tag{4.63}$$

Observe that the transfer functions $t_i(s)$ of $T_D(s)$ have relative degree equal to the corresponding columns' degrees of $\xi_l(s)$, which guarantees the existence of a proper controller that achieves the desired closed-loop dynamics. The decoupling controller is then

$$Q(s) = P^{-1}(s)T_D(s) \tag{4.64}$$

$$= \frac{(s+2)(s+0,5)^2}{(s^2+3s+4)(s^2+2s+7)} \begin{bmatrix} -8(s-2) & -16(s+0,5) \\ 24 & -8(s-2) \end{bmatrix}$$

As the IMC controller is stable, the feedback system is internally stable.

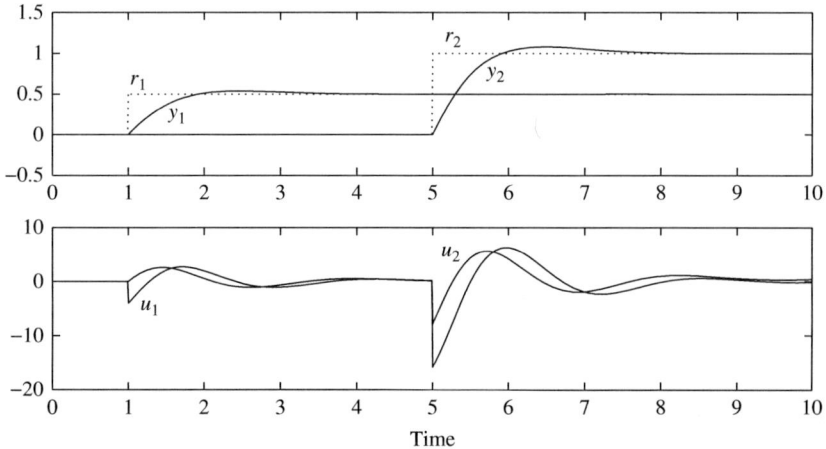

Figure 4.5 Dynamic decoupling of the MP system described by (4.61)

Figure 4.5 presents the output responses in each channel and the corresponding control actions for the multivariable closed-loop system using controller (4.64) when it is excited with step reference signals. As can be observed, both the full decoupling and the desired closed-loop dynamics are achieved.

4.3.2   *Non-minimum phase systems*

As could be expected, finding an approximate inverse for the case of NMP processes is something more involved because $P^{-1}(s)$ has now unstable poles. We must therefore modify $Q(s)$ so as to be stable but to preserve its diagonalising properties also.

To realise this aim, we need to add an additional compensator $D_z(s)$ to the controller $Q(s)$ obtained for dynamic decoupling of MP processes [49]. Let us see how it should be and how to compute it.

Assume that $\Lambda_I(s) = \xi_I(s)P(s)$ has the following description in the state space:

$$\dot{x}(t) = \bar{A}x(t) + \bar{B}u \tag{4.65}$$

$$y = \bar{C}x(t) + \bar{D}u \tag{4.66}$$

where $\bar{A}, \bar{B}, \bar{C}$ and $\bar{D}$ $(det(\bar{D}) \neq 0)$ are the 'A', 'B', 'C' and 'D' matrices of the plant modified by the interactor matrix compensation. The inverse of (4.65) and (4.66) can be obtained by simply reversing the roles of inputs and outputs, which gives rise to the following realisation of $\Lambda_I^{-1}(s)$:

$$\dot{x}(t) = \bar{A}x(t) + \bar{B}\bar{D}^{-1}(y - \bar{C}x(t)) \tag{4.67}$$

$$\hat{u} = \bar{D}^{-1}(y - \bar{C}x(t)) \tag{4.68}$$

with $\hat{u}$ being the inverse output ($\hat{u}$ is an estimation of the plant input u). Regrouping terms and denoting $\tilde{u} = y$, the state-space representation given in (4.67) and (4.68) can be written as

$$\dot{x}(t) = A_\lambda x(t) + B_\lambda \tilde{u} \tag{4.69}$$

$$\hat{u} = C_\lambda x(t) + D_\lambda \tilde{u} \tag{4.70}$$

where

$$A_\lambda = \bar{A} - \bar{B}\bar{D}^{-1}\bar{C} \tag{4.71}$$

$$B_\lambda = \bar{B}\bar{D}^{-1} \tag{4.72}$$

$$C_\lambda = -\bar{D}^{-1}\bar{C} \tag{4.73}$$

$$D_\lambda = \bar{D}^{-1} \tag{4.74}$$

Note that the matrix A_λ in (4.69) has eigenvalues with positive real part because of the RHP zeros of $P(s)$. Thus, the compensator $D_z(s)$ to be added will be the resultant transfer matrix from performing a state feedback stabilisation (with gain k_i) of each subsystem of (4.69) and (4.70), with the ith component of the vector $\tilde{u}$ as input. These subsystems can be represented by

$$\dot{x}_i(t) = A_i x_i(t) + b_i \tilde{u}_i(t) \tag{4.75}$$

$$v_i(t) = C_i x_i(t) + d_i \tilde{u}_i(t) \tag{4.76}$$

where $v_i(t) \in \mathbb{R}^m$, $\tilde{u}_i(t) \in \mathbb{R}$ and (A_i, b_i, C_i, d_i) is a minimal realisation of $(A_\lambda, B_\lambda e_i, C_\lambda, D_\lambda e_i)$, with e_i being the ith column of the identity matrix I_m.

Hence, $D_z(s)$ is given by

$$D_z(s) = diag\left(\left[1 + k_i(sI - A_i)^{-1} b_i \right]^{-1} \right) \tag{4.77}$$

A stable approximate inverse of $P(s)$ is then given by the following IMC controller:

$$Q(s) = \Lambda_I^{-1}(s)\xi_I(s)D_z(s)T_D(s) \tag{4.78}$$

This is equivalent to

$$Q(s) = P^{-1}(s) \, diag\left(\left[1 + k_i(sI - A_i)^{-1} b_i \right]^{-1} \right)$$
$$\cdot diag(t_1(s), t_2(s), \ldots, t_m(s)) \tag{4.79}$$

From the previous design, the complementary sensitivity matrix $T(s)$ results:

$$T(s) = diag\left(\left[1 + k_i(sI - A_i)^{-1} b_i \right]^{-1} t_i(s) \right) \tag{4.80}$$

Since the RHP zeros are unstable eigenvalues of the matrices A_i, they will appear unaltered in the matrix $T(s)$, which is necessary for the internal stability of the feedback system. In fact, although $D_z(s)$ stabilises $\Lambda_r^{-1}(s)$ – which could be seen as a unstable pole/zero cancellation – these transfer matrices have only sense in order to compute the controller $Q(s)$, but they do not constitute a physical subsystem of the control loop. Furthermore, the RHP zeros will spread among the different channels of the decoupled system depending on their output directions. This is an inherent limitation of NMP system decoupling, which we will further analyse in this chapter and which will be a problem to be addressed in the remainder of the book.

Example 4.6: Consider a 2×2 plant with the following nominal model

$$P(s) = \frac{1}{(s+1)(s+2)} \begin{bmatrix} s+1 & 2 \\ 1 & 1 \end{bmatrix} \tag{4.81}$$

This is an NMP plant with a transmission zero at $s = +1$. Consequently, we must follow the procedure just described in order to synthesise a controller that decouples the closed-loop system. We first compute the left ($\xi_l(s)$) interactor of $P(s)$ as explained in Section 4.2.3, with the roots of interactor polynomials at $\alpha = -1$ (see Remark 4.3). Hence, by simple inspection of (4.81)

$$
\xi_l(s) = \begin{bmatrix} s+1 & 0 \\ 0 & (s+1)^2 \end{bmatrix} \tag{4.82}
$$

Note that in this case, the right and left interactors coincide ($\xi_l(s) = \xi_r(s)$) . If we select – without loss of generality – $\xi_l(s)$ to perform the design, we have

$$
\Lambda_l^{-1}(s) = [\xi_l(s)P(s)]^{-1} = \frac{s+2}{s-1} \begin{bmatrix} 1 & -2 \\ \dfrac{-1}{s+1} & 1 \end{bmatrix} \tag{4.83}
$$

which is unstable as expected. One should then compute the compensator $D_z(s)$ in the state space as previously described. $\Lambda_l^{-1}(s)$ has the following realisation:

$$
\dot{x}(t) = \begin{bmatrix} -0.69 & 0.63 \\ 0.84 & 0.69 \end{bmatrix} x(t) + \begin{bmatrix} -0.75 & 0.63 \\ -0.63 & 1.69 \end{bmatrix} \tilde{u} \tag{4.84}
$$

$$
\hat{u} = \begin{bmatrix} -1.5 & -3 \\ -0.22 & 1.86 \end{bmatrix} x(t) + \begin{bmatrix} 1 & -2 \\ 0 & 1 \end{bmatrix} \tilde{u} \tag{4.85}
$$

Thus, the subsystem with input $\tilde{u}_1$ (subsystem 1) has a minimal realisation of the form given in (4.75) and (4.76), with matrices

$$
A_1 = \begin{bmatrix} -0.69 & 0.63 \\ 0.84 & 0.69 \end{bmatrix}, \quad b_1 = \begin{bmatrix} -0.75 \\ -0.63 \end{bmatrix}
$$
$$
C_1 = \begin{bmatrix} -1.5 & -3 \\ -0.22 & 1.86 \end{bmatrix}, \quad d_1 = \begin{bmatrix} 1 \\ 0 \end{bmatrix} \tag{4.86}
$$

while the matrices corresponding to the state-space description of the subsystem with input $\tilde{u}_2$ (subsystem 2) are

$$A_2 = 1, \quad b_2 = 1.8$$

$$C_2 = \begin{bmatrix} -3.33 \\ 1.67 \end{bmatrix}, \quad d_2 = \begin{bmatrix} -2 \\ 1 \end{bmatrix} \tag{4.87}$$

Now, both subsystems are stabilised using state feedback. The computation of the gains k_1 and k_2 can be carried out with any pole assignment method. As the eigenvalues of subsystem 1 are $(\lambda_1, \lambda_2) = (1, -1)$, we choose $k_1 = [-5.5 \; -11]$ to shift them to $(\lambda'_1, \lambda'_2) = (-10, -1)$. Subsystem 2 has a unique eigenvalue $\lambda = 1$, which is moved to $\lambda' = -1$ with $k_2 = 1.11$.

In this way, the compensator $D_z(s)$ in (4.77) results

$$D_z(s) = \begin{bmatrix} \dfrac{s-1}{s+10} & 0 \\ 0 & \dfrac{s-1}{s+10} \end{bmatrix} \tag{4.88}$$

What follows then is to choose $T_D(s) = diag(t_1(s), t_2(s))$ according to the desired closed-loop dynamics. We recall that the transfer functions $t_i(s)$ must have relative degree equal to the degree of the interactor columns in (4.82). In addition, to have zero steady-state error, the DC-gain of $t_i(s)$ must be such that $T(0) = I_2$. Assuming a desired time constant of 1 [time unit] for the closed loop, we take

$$T_D(s) = \begin{bmatrix} \dfrac{-(s+10)}{s^2 + 2s + 1} & 0 \\ 0 & \dfrac{-1}{s^2 + 2s + 1} \end{bmatrix} \tag{4.89}$$

Then, from (4.78), the IMC decoupling controller is

$$Q(s) = \frac{s+2}{s+1} \begin{bmatrix} -1 & \dfrac{2}{s+1} \\ 1 & -1 \end{bmatrix} \tag{4.90}$$

and the closed-loop transfer matrix from the reference to the output vector results:

$$T(s) = \begin{bmatrix} \dfrac{-s+1}{s^2 + 2s + 1} & 0 \\ 0 & \dfrac{-s+1}{s^3 + 3s^2 + 3s + 1} \end{bmatrix} \tag{4.91}$$

As can be appreciated, although the complementary sensitivity is diagonal as aimed, it has an RHP zero in each channel at $s = 1$.

The response of the closed-loop system with the controller $Q(s)$ to square reference signals of frequency $1/60$ and $1/40$ [time units]$^{-1}$ is shown in Figure 4.6. It verifies the dynamic decoupling of the NMP system and shows how the RHP zero affects both channels with inverse responses, which, as was already mentioned, is the price to be paid when NMP plants are diagonally decoupled.

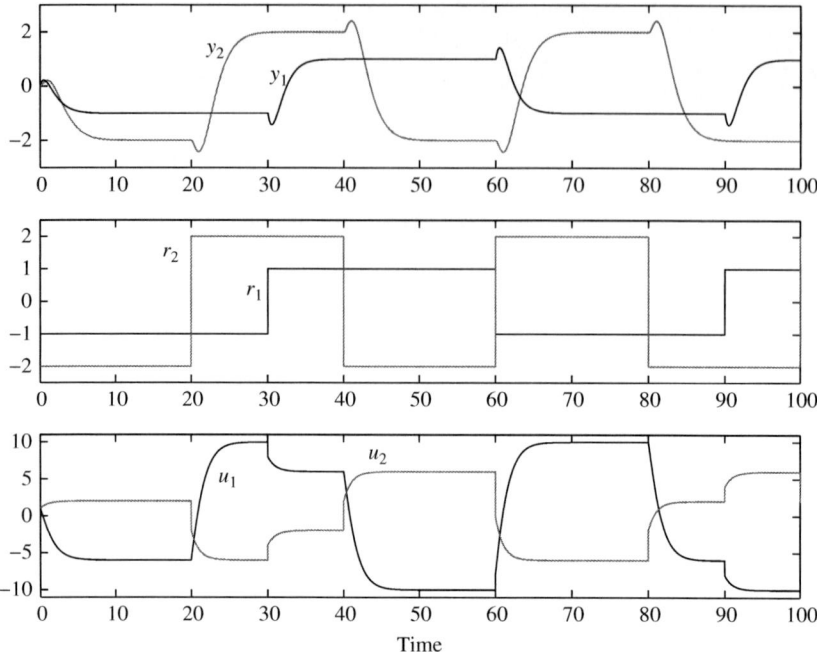

Figure 4.6   Controlled variables (y_i), reference signals (r_i) and control actions (u_i) of the NMP system with controller (see (4.90))

4.3.3   Unstable systems

When the multivariable system is open-loop unstable, the affine parameterisation presented in Section 4.2 cannot be used for decoupling purposes. Instead, there exist two degree-of-freedom strategies based on parameterisations of non-necessarily stable multivariable systems or on the prefiltering of reference signals [48,80,81]. Herein, we will only present the latter since it allows using the strategies studied in Sections 4.3.1 and 4.3.2. Further, it is easier to implement than the techniques based on unstable system parameterisations.

The prefiltering technique simply consists of stabilising first the open-loop unstable system with a proper controller $C(s)$, and to introduce then a pre-compensation of the input signal to the control loop, which attains the input–output decoupling of the stabilised system. The basic idea is represented in Figure 4.7.

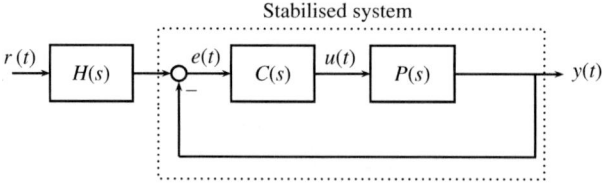

Figure 4.7   Prefiltering of the reference vector for the decoupling of unstable systems

Given the transfer matrix model $P(s)$ of an unstable system, the input–output closed-loop relationship can be written as

$$T(s) = [I + P(s)C(s)]^{-1}P(s)C(s)H(s) \qquad (4.92)$$

Then, to achieve the dynamic decoupling of the system, one has just to design $H(s)$ in the same way as explained for the $Q(s)$ design in the case of stable systems. That is, $H(s)$ must be a right diagonalising precompensator of the stable transfer matrix $[I + P(s)C(s)]^{-1}P(s)C(s)$.

The main drawback of decoupling by reference prefiltering is that with this strategy, one can only decouple the controlled variables from the reference signals, but not from other input signals such as output disturbances.

Remark 4.5: *Although Figure 4.7 shows an open-loop configuration of $H(s)$ compensation, it can obviously be implemented in a closed-loop IMC scheme as the ones considered in previous sections.*

4.4 Performance limitations in non-minimum phase systems

Section 4.3.2 revealed that the presence of multivariable RHP zeros makes dynamic decoupling more difficult. Moreover, the resultant transfer matrix in (4.80) suggests that a unique RHP zero may affect multiple channels of the closed loop when it is dynamically decoupled. Let us see now the reasons of this undesired RHP zero spreading.

Consider a nominal model of a plant, $P(s) \in \mathbb{R}^{m \times m}(s)$, having an RHP zero located at $s = z_0$, $z_0 > 0$, with output direction $h^{\mathrm{T}} = [h_1\ h_2\ \dots\ h_m]$. Thus,

$$h^{\mathrm{T}}P(z_0) = 0 \in \mathbb{R}^m \qquad (4.93)$$

Assume now that a decoupling controller $C(s)$ is designed. Then, both the complementary sensitivity $T(s)$ and the sensitivity $S(s) = I - T(s)$ will be diagonal matrices. This means that the open-loop transfer matrix

$$L(s) \overset{\Delta}{=} P(s)C(s) \qquad (4.94)$$

will also be diagonal. That is,

$$L(s) = diag(l_{11}(s), l_{22}(s), \dots, l_{mm}(s)) \qquad (4.95)$$

$$\begin{aligned} S(s) &= [I + L(s)]^{-1} \\ &= diag(s_{11}(s), s_{22}(s), \dots, s_{mm}(s)) \end{aligned} \qquad (4.96)$$

$$\begin{aligned} T(s) &= L(s)[I + L(s)]^{-1} \\ &= diag(t_{11}(s), t_{22}(s), \dots, t_{mm}(s)) \end{aligned} \qquad (4.97)$$

According to (4.93) and the definition of $L(s)$ given in (4.94), the following equality must be satisfied:

$$h^\mathsf{T} T(z_o) = [h_1 t_{11}(z_o) \, h_2 t_{22}(z_o) \, \ldots \, h_m t_{mm}(z_o)] = 0 \in \mathbb{R}^m \tag{4.98}$$

So, $h_i t_{ii}(z_0) = 0$ for $i = 1, \ldots, m$. This means that $t_{ii}(z_0) = 0$ for every value of i in which the corresponding component h_i is non-zero.

One can therefore conclude that the RHP transmission zeros will necessarily spread when decoupling the system if their output direction has more than one non-zero entry. This spreading can only be avoided in those particular cases in which the output directions of RHP zeros are canonical, i.e. they have only one non-zero component.

It is worth pointing out that the interpolation constraint (4.98) is a direct implication of the internal stability requirement.

Example 4.7: Consider a plant having the following model

$$P(s) = \frac{1}{(s+1)(s+3)} \begin{bmatrix} s+3 & 4 \\ 2 & 2 \end{bmatrix} \tag{4.99}$$

which has a single RHP zero at $s = +1$ with unitary geometric multiplicity and output direction $h^\mathsf{T} = [1 \; -2]$.

Assume that the following diagonal complementary sensitivity is aimed:

$$T_D(s) = P(s)Q(s) = \begin{bmatrix} t_{11}(s) & 0 \\ 0 & t_{22}(s) \end{bmatrix} \tag{4.100}$$

The plant model inverse is given by

$$P^{-1}(s) = \frac{(s+1)(s+3)}{2(s-1)} \begin{bmatrix} 2 & -4 \\ -2 & s+3 \end{bmatrix} \tag{4.101}$$

which verifies the presence of the zero at $s = 1$ in (4.99). Note that the inverse (4.101) is clearly improper, and therefore not implementable. This is the reason why we have studied how to compute approximate proper inverses of $P(s)$. From (4.100), the IMC controller must have the following form:

$$Q(s) = P^{-1}(s) \begin{bmatrix} t_{11}(s) & 0 \\ 0 & t_{22}(s) \end{bmatrix} \tag{4.102}$$

Since the controller $Q(s)$ must be stable for the internal stability of the feedback system, both $t_{11}(s)$ and $t_{22}(s)$ must cancel at $s = 1$. Therefore, an inverse response characteristic will appear in both closed-loop channels, which is consistent with the previous analysis, because the two entries of the output direction $h^\mathsf{T} = [1 \; -2]$ are non-zero.

Consider now this different model:

$$P'(s) = \frac{1}{(s+1)(s+3)} \begin{bmatrix} 1 & 2 \\ s-1 & s-1 \end{bmatrix} \tag{4.103}$$

which also has an RHP zero at $s = +1$ with unitary geometric multiplicity, but with canonical output direction $h^T = [0 \ 1]$.

In this case, the inverse of the plant is

$$P'^{-1}(s) = \begin{bmatrix} -(s+1)(s+3) & \frac{2(s+1)(s+3)}{s-1} \\ (s+1)(s+3) & \frac{-(s+1)(s+3)}{s-1} \end{bmatrix} \tag{4.104}$$

and thus the IMC controller must be

$$Q(s) = \begin{bmatrix} -(s+1)(s+3)t_{11}(s) & \frac{2(s+1)(s+3)}{s-1}t_{22}(s) \\ (s+1)(s+3)t_{11}(s) & \frac{-(s+1)(s+3)}{s-1}t_{22}(s) \end{bmatrix} \tag{4.105}$$

As can be observed, for $Q(s)$ to be stable it is only required that $t_{22}(s)$ has a zero at $s = 1$, but not $t_{11}(s)$. So, for this plant in which the RHP zero has a canonical output direction, the NMP affects only one closed-loop channel.

Apart from producing the RHP zero spreading, the decoupling of NMP systems tightens up the fundamental limitations of multivariable feedback system design (provided the RHP zero directions are not canonical). The expressions that quantify these costs of NMP decoupling can be found in Reference 110. In the time domain, they constitute a constraint over an error signal integral, while in the frequency domain, the limitations arise from the so-called Poisson integral. The latter gives rise to lower bounds on the ∞-norm of the sensitivity function of the system, which becomes harder as the decoupling bandwidth increases.

Chapter 5

Constrained dynamic decoupling

In Chapter 4, we described a methodology to achieve the closed-loop dynamic decoupling of multivariable systems. Although some difficulties arising from right-half plane (RHP) zeros were pointed out, the decoupling design implicitly assumed unconstrained systems and centralised controllers, as in general do the great majority of the existing techniques. Hereinafter (Chapters 5–7), we will take advantage of the sliding mode reference conditioning (SMRC) features to improve the closed-loop decoupling in the presence of either physical (actuator saturations), structural (decentralised controllers) or dynamic (non-minimum phase (NMP) plants) limitations.

When the unavoidable *physical limits* of the real actuators are taken into account, the activation of any of them produces a change in the direction of the plant input with respect to the controller output and, as a consequence, the decoupling obtained for linear operation is lost. In this chapter, we first illustrate the *control directionality* problem briefly introduced in Chapter 1, and we then present a compensation method using SMRC ideas to preserve the closed-loop dynamic decoupling in presence of input constraints.

5.1   Introduction

Several control techniques exist in order to design controllers that achieve full dynamic decoupling when ideal actuators are considered, like the ones described or cited in Chapter 4. However, the problem becomes considerably harder if physical limits of the actuators are taken into account. In fact, multiple input saturation changes the amplitude and the direction of the control signal that is necessary to achieve dynamic decoupling. Hence, in addition to the known problem of windup, the control directionality problem appears, which brings about the loss of the decoupling obtained for the ideal case.

There have been many efforts to preserve control directionality in constrained multivariable systems. Hanus and Kinnaert [54] first proposed to modify the reference vector. Therein, an artificial non-linearity placed just before the real non-linearity is designed in such a way that the modified reference remains as close as possible to the original one, under some criterion. Afterwards, Walgama and Sternby [142] integrated the ideas of Hanus and Kinnaert with a generalisation of the conditioning technique by introducing a filtered set point. A posterior contribution worthy of mention was made by Peng *et al.* [95], where a parameterisation of anti-windup compensators and an optimal design were addressed in a simple manner. Other techniques using tools such as Linear Matrix Inequalities (LMI), Linear Parameter-Varying (LPV) and Internal Model Control (IMC) were also applied to this topic [90,151,159].

However, the preservation of control directionality is a necessary but not sufficient condition to preserve dynamic decoupling in the presence of input constraints. Most of the optimal design methods successfully avoid the change of control directionality by conditioning the whole reference vector. In this way, although they solve the problem that originally caused the loss of decoupling, the methodology used may – in lesser degree – also affect the decoupling of the system. In fact, when a single reference changes, the simultaneous correction of the whole reference vector may lead to shape the unchanged references, thus producing transient effects in controlled variables that should not change. Hence, when the process to be controlled allows reaching the operating points by successive changes of individual reference components, which is very common in industrial processes, an improved degree of decoupling is achievable. We will therefore focus on this way of operation, which is also taken into account in Reference 49.

In this chapter, we present a technique to preserve dynamic decoupling of constrained multivariable processes, ensuring that the controlled variables whose set points did not change remain unaffected. The algorithm is based on SMRC ideas, and it can be combined with most decoupling techniques valid for linear operation.

Note that despite the performance costs already claimed in Chapter 4, we will here consider full dynamic decoupling as the main control objective. Chapters 6 and 7 will propose alternative strategies to relax some drawbacks of diagonal decoupling.

5.2   Control directionality changes

Similar to what happens with single-input single-output (SISO) systems, the presence of input constraints in multiple-input multiple-output (MIMO) systems leads to an inconsistency between the controller states (or outputs) and the plant inputs. This inconsistency produces a serious degradation of closed-loop performance, which can be interpreted as a windup-related problem.

Furthermore, since in multivariable systems saturation can occur independently in each channel, an additional problem known as *control directionality* comes out, characteristic of MIMO systems. As we have seen in Chapter 4, an accurate coupling between the process and the controller is required to decouple a multivariable control system. Indeed, the controller must include an approximation of the plant inverse in order to avoid closed-loop interactions. What occurs in the presence of multiple actuators is that the individual action of each saturating element affects the required coupling between the controller and the plant, modifying the necessary direction of the control vector to attain decoupling. Consequently, the decoupling obtained for the linear operation is lost. The next example illustrates the closed-loop effects of the control directionality change.

Example 5.1: Consider a problem that has been used to demonstrate the performance of many constrained control strategies, originally presented in Reference 16. The process to be controlled is given by

$$P(s) = \frac{1}{10s + 1} \begin{bmatrix} 4 & -5 \\ -3 & 4 \end{bmatrix} \tag{5.1}$$

and is dynamically decoupled by means of the controller:

$$C(s) = \frac{10s + 1}{s} \begin{bmatrix} 4 & 5 \\ 3 & 4 \end{bmatrix} \tag{5.2}$$

To facilitate the connection between this controller and the methodologies presented in Chapter 4, it is important to note that this controller leads to the following closed-loop transfer matrix:

$$T(s) = T_D(s) = \begin{bmatrix} \frac{1}{s+1} & 0 \\ 0 & \frac{1}{s+1} \end{bmatrix} \tag{5.3}$$

whose elements have relative degree equal to the degree of the corresponding columns of the left interactor of $P(s)$:

$$\xi_l(s) = \begin{bmatrix} s & 0 \\ 0 & s \end{bmatrix} \tag{5.4}$$

According to (4.27), $C(s)$ corresponds to the following IMC controller

$$Q(s) = P^{-1}(s)T_D(s) = \frac{10s + 1}{s + 1} \begin{bmatrix} 4 & 5 \\ 3 & 4 \end{bmatrix} \tag{5.5}$$

which is biproper and stable, and therefore the closed-loop system is internally stable.

The control system with controller (5.2) is now excited by step references filtered above a cut-off frequency of $\omega_c = 1$ (i.e. a filter with the same bandwidth as the closed loop was added). The purpose of this filtering is not to avoid the problems caused by input saturation but to allow a better visualisation of these effects in the plane (u_1, u_2) (see Figure 5.3).

The results obtained for the ideal unconstrained system are shown in Figure 5.1. As can be observed, controller (5.2) achieves full dynamic decoupling of the system during its linear operation. The controller outputs u coincide in this case with the plant inputs $\hat{u}$, as reveal the curves of the lower part of the figure.

If we now add a non-linear saturating element representing the amplitude limitation of each actuator, the response is seriously degraded. We take for this example the limit $\bar{u} = 15$ for both actuators. Figure 5.2 shows that the system remains decoupled when facing a reference step for y_1. This is because the control

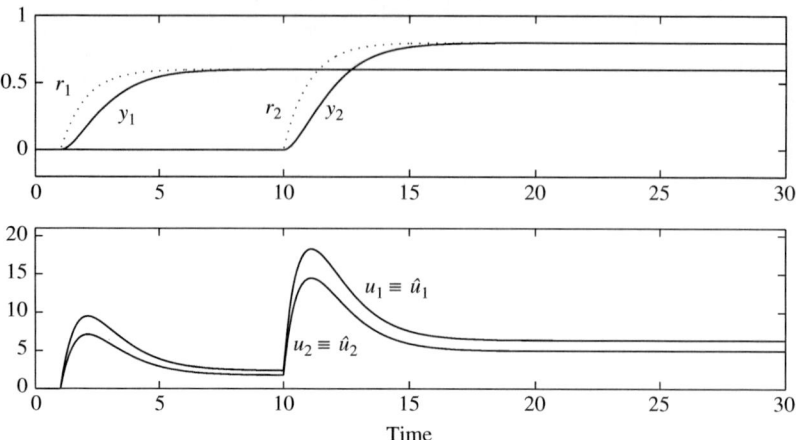

Figure 5.1   Evolution of the controlled variables y and the control actions u for the unconstrained system

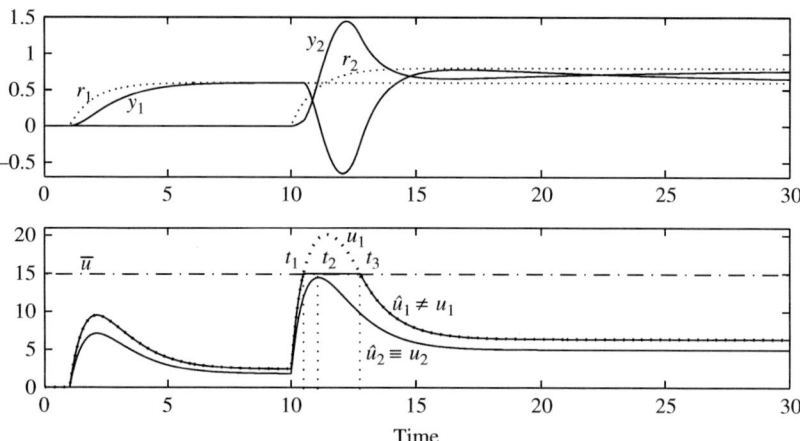

Figure 5.2   Evolution of the controlled variables y and the control actions u for the system with input saturation

vector does not reach its limits, hence preserving its direction. On the contrary, extremely poor closed-loop responses are evidenced when a set-point change is applied to y_2, because of the first actuator saturation between t_1 and t_3 ($u \neq \hat{u}$).

The loss of decoupling in Figure 5.2 is caused by the aforementioned control directionality change, as show the trajectories of the control action in the plane (u_1, u_2) plotted in Figure 5.3. The angle between the plant input vector $\hat{u}(t)$ and the controller output $u(t)$, different from zero between t_1 and t_3 (see also Figure 5.2), indicates the control directionality change. In Figure 5.3, the dashed arrows show the directional gap between $\hat{u}(t)$ and $u(t)$ for the time instant t_2 ($t_1 < t_2 < t_3$).

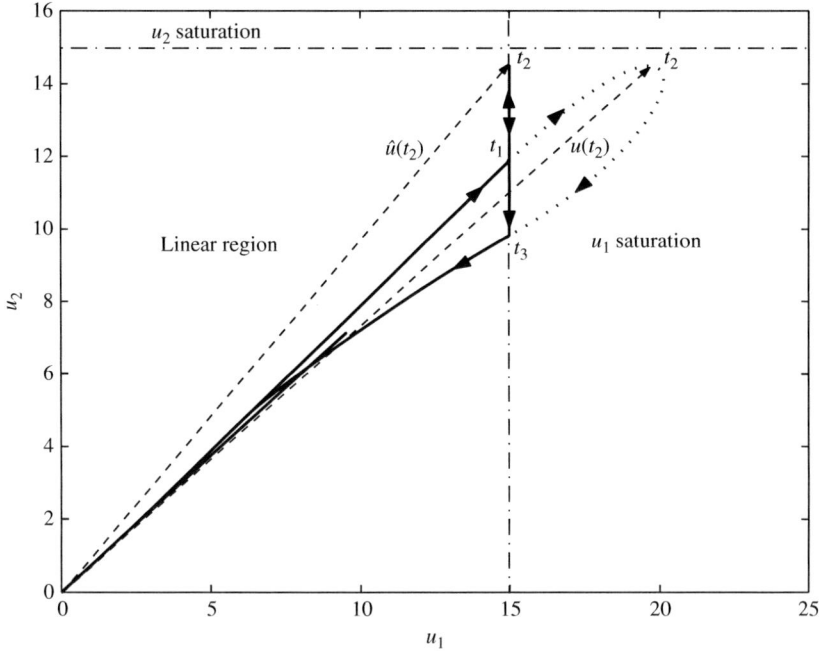

Figure 5.3   Control directionality change

5.3   Dynamic decoupling preservation by means of SMRC

On the basis of the SMRC ideas, a method to preserve the dynamic decoupling of multivariable systems in the presence of input constraints is described here [40]. The general case of proper (biproper or strictly proper) controllers is considered.

5.3.1   Method formulation

Figure 5.4 represents the proposed MIMO control scheme, where the main unitary feedback control loop and the SMRC conditioning loop can be easily distinguished.

In the main control loop, $P(s) \in \mathbb{R}^{m \times m}(s)$ represents a stable process under control, which could be minimum-phase or not. K_a are m actuators with saturation, whereas $C(s) \in \mathbb{R}^{m \times m}(s)$ is a proper centralised controller designed to achieve closed-loop diagonal decoupling during linear operation of the actuators K_a. $F(s)$ represents a first-order linear filter in each channel. The signals $\mathbf{r}, \mathbf{r_f}, \mathbf{e}, \mathbf{u}, \hat{\mathbf{u}}$ and $\mathbf{y}$ are vectors of m scalar functions of time.[1]

Although any type of non-linearity could have been considered, for the sake of simplicity, we model the non-linearity introduced by the ith actuator as an individual amplitude saturation:

---

[1] In this chapter, we will denote signal vectors with bold typeface, so as to distinguish them from scalar variables of Chapter 2.

$$k_{a_i} : \begin{cases} \hat{u}_i = \bar{u}_i & \text{if} \quad u_i > \bar{u}_i \\ \hat{u}_i = u_i & \text{if} \quad \underline{u}_i \leq u_i \leq \bar{u}_i \\ \hat{u}_i = \underline{u}_i & \text{if} \quad u_i < \underline{u}_i \end{cases} \tag{5.6}$$

with $i = 1, \ldots, m$, $\underline{u}_i$ and $\bar{u}_i$ being the corresponding elements of $\underline{\mathbf{u}} \in \mathbb{R}^m$ and $\bar{\mathbf{u}} \in \mathbb{R}^m$, respectively.

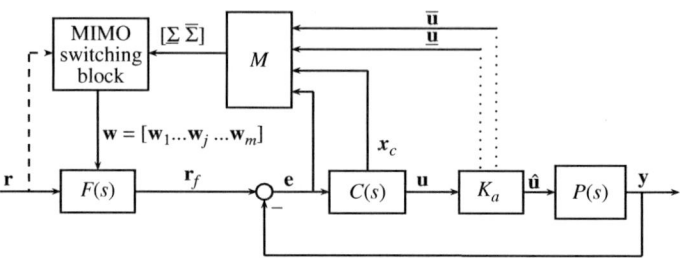

Figure 5.4   SMRC loop to preserve dynamic decoupling of MIMO systems

The SMRC loop defines for this MIMO problem two sliding surfaces for each one of the m^2 transfer functions of the controller. This requires defining some new elements in the compensation loop.

The block M represents a constant matrix operator, which generates two switching vectors, $\bar{\Sigma}$ and $\underline{\Sigma} \in \mathbb{R}^{m^2}$, the former comprising the switching functions associated with the upper limits of the actuators and the latter comprising the ones corresponding to the lower bounds. Within $\tilde{\Sigma} = [\tilde{\sigma}_1^T \ldots \tilde{\sigma}_j^T \ldots \tilde{\sigma}_m^T]^T$ (recall $\tilde{\star}$ stands for either $\bar{\star}$ or $\underline{\star}$), each $\tilde{\sigma}_j = [\tilde{\sigma}_{1j} \ldots \tilde{\sigma}_{ij} \ldots \tilde{\sigma}_{mj}]^T$ is a vector composed of the switching functions associated with the jth reference conditioning. Note that each $\tilde{\sigma}_{ij}$ comes from a different actuator K_{a_i}.

To preserve dynamic decoupling, the discontinuous vector $\mathbf{w} = [\mathbf{w}_1^T \ldots \mathbf{w}_j^T \ldots \mathbf{w}_m^T]^T$, wherein each $\mathbf{w}_j = [w_{1j} \ldots w_{ij} \ldots w_{mj}]^T$ comprises the discontinuous signals that shape the jth reference, is governed by the following switching law (implemented in the *MIMO switching block*):

$$\begin{cases} w_{ij} = w_{ij}^- & \text{if} \quad \bar{\sigma}_{ij} < 0 \\ w_{ij} = w_{ij}^+ & \text{if} \quad \underline{\sigma}_{ij} > 0, \quad i = 1, \ldots, m; \quad j = j_0 \\ w_{ij} = 0 & \text{otherwise} \end{cases} \tag{5.7}$$

while $w_{ij} = 0$ if $j \neq j_0$, r_{j_0} being the reference that has been changed. Observe that each of the signals $w_{ij_0} \in \mathbf{w}_{j_0}$ considered in (5.7) corresponds to one of the m actuators that can saturate.

Remark 5.1: *Although stable open-loop systems have been assumed, the approach does not present a priori limitations for being applied to unstable systems subjected to constraints. In such a case, the system can only be locally stabilised for a limited set of initial conditions. This is a structural limitation of every unstable constrained* *system and not a particular limitation of the proposed approach. As with any other control strategy, an invariant set of the state space should be derived for a suitable range of reference values, in which control signals limits and stability are ensured.*

5.3.2   Sliding surfaces design

As we have seen in Section 2.5, if one aims to limit an output with relative degree greater than one, the switching function must include as many derivatives of that output as necessary for having unitary relative degree with respect to the discontinuous action.

Effectively, for strictly proper transfer functions of the controller, the corresponding version of (2.5) has relative degree greater than one with respect to w_{ij}. Thus, the sliding functions will have to include other controller states apart from its outputs to enable the establishment of sliding regimes. Although this makes the sliding functions a bit more complex, it also provides degrees of freedom that may be used to control the rate of approach to the saturation limits. Hence, the following sliding functions are defined:

$$\tilde{\sigma}_{ij} = \tilde{u}_i - u_i, \quad \text{if} \quad \rho_{ij} = 0 \tag{5.8}$$

$$\tilde{\sigma}_{ij} = \tilde{u}_i - u_i - \sum_{\alpha=1}^{\rho_{ij}} k_\alpha^{ij} u_i^{(\alpha)}, \quad \text{if} \quad \rho_{ij} \geq 1 \quad i,j = 1, ..., m \tag{5.9}$$

where ρ_{ij} is the relative degree of the transfer function between the controller output u_i and input e_j, $u_i^{(\alpha)}$ is the derivative of order α of u_i and k_α^{ij} are constant gains. Note that the inclusion of $u_i^{(\rho)}$ guarantees in this case that the sliding functions are of relative degree one with respect to w_j, and that $u_i^{(\alpha)}$ can be obtained as a linear combination of the controller states and inputs (which are obviously accessible) in the constant matrix operator M.

Similar to the previous SMRC schemes operation, (5.7)–(5.9) state that SMRC will be transiently activated on the surface $\tilde{\sigma}_{ij} = 0$ to shape the reference signal and prevent controller outputs from crossing their limits.

5.3.3   SMRC dynamics

Here, we will give an alternative SMRC dynamic analysis with respect to the one given in Chapter 2. As the plant $P(s)$ in Figure 5.4 was assumed to be stable, the transfer function from **r** to **u** ($Q(s)$) not only has the same relative degree structure of controller $C(s)$ but also is minimum phase (MP) if and only if $C(s)$ is MP as well. Then, we will consider the controller $C(s)$ as the constrained subsystem and the output feedback y as a bounded disturbance during SMRC operation. It is important to recall that this analysis is only valid provided $P(s)$ is stable.

The filter $F(s)$ may be represented in state space as

$$F(s) : \begin{cases} \dot{\mathbf{x}}_f = A_f \mathbf{x}_f + B_f \mathbf{r} + B_w \mathbf{w} \\ \mathbf{r}_f = C_f \mathbf{x}_f \end{cases} \tag{5.10}$$

where $A_f = -C_f = \lambda_f \cdot I_m$ (λ_f eigenvalue), $B_f = I_m$ and $B_w \in \mathbb{R}^{m \times m^2}$ is block diagonal (blocks of $1 \times m$).

Consider also the following column realisation [18] of the controller:

$$C(s) : \begin{cases} \dot{\mathbf{x}}_c = A_c \mathbf{x}_c + B_c \mathbf{e} \\ \mathbf{u} = C_c \mathbf{x}_c + D_c \mathbf{e} \end{cases} \tag{5.11}$$

in which $A_c = diag(A_j)$, $B_c = diag(\mathbf{b}_j)$, $C_c = [C_1 \ldots C_m]$ and $D_c = [\mathbf{d}_1 \ldots \mathbf{d}_m]$, with $j = 1, \ldots, m$. A_j and C_j are matrices of $r_{d_j} \times r_{d_j}$ and $m \times r_{d_j}$, respectively, and r_{d_j} is the degree of the least common denominator of the transfer functions of the jth column of $C(s)$. Also, $\mathbf{b}_j$ and $\mathbf{d}_j$ are column vectors of r_{d_j} and m elements, respectively. Hence, $(A_j, \mathbf{b}_j, C_j, \mathbf{d}_j)$ is a realisation of the transfer vector between the error e_j and the controller outputs $\mathbf{u}$. Picking out the ith row of C_j and the ith element of $\mathbf{d}_j$ (called $\mathbf{c}_{ij}^T$ and d_{ij}, respectively), it results in a realisation of the transfer function between e_j and the output u_i:

$$c_{ij}(s) : \begin{cases} \dot{\mathbf{x}}_{c_j} = A_j \mathbf{x}_{c_j} + \mathbf{b}_j e_j \\ u_i = \mathbf{c}_{ij}^T \mathbf{x}_{c_j} + d_{ij} e_j \end{cases} \tag{5.12}$$

As we argued in the introduction, individual changes of the reference components are considered because this way of operation is very common in decoupled designs and it allows achieving an improved degree of decoupling in the presence of input saturation. Then, and according to (5.7), the SMRC loop does always shape the reference that was last changed, which we will call r_j from now on instead of r_{j_0} to simplify notation. In this way, each time the system is about to reach saturation in the ith component of the control vector $\mathbf{u}$, the sliding mode (SM) compensation will operate over the controller transfer function $c_{ij}(s)$, as shown in Figure 5.5. The other components of the error $\mathbf{e}$ do not affect the SMRC loop because they remain unchanged.

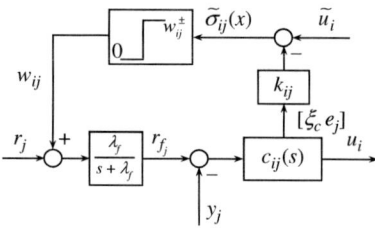

Figure 5.5   Active SMRC loop when a limit $\tilde{u}_i$ is reached because of a change in r_j

Therefore, the dynamic analysis is analogous to the one given in Chapter 2. In fact, the open-loop dynamics of the SMRC compensation is given from (5.10) and (5.12) by

$$\begin{bmatrix} \dot{\mathbf{x}}_{c_j} \\ \dot{e}_j \end{bmatrix} = \begin{bmatrix} A_j & \mathbf{b}_j \\ 0 & 0 \end{bmatrix} \begin{bmatrix} \mathbf{x}_{c_j} \\ e_j \end{bmatrix} + \begin{bmatrix} 0 \\ \mathbf{c}_{f_j}^T A_f \mathbf{x}_f + \mathbf{c}_{f_j}^T B_f \mathbf{r} - \dot{y}_j \end{bmatrix} + \begin{bmatrix} 0 \\ \mathbf{c}_{f_j}^T B_w \mathbf{w} \end{bmatrix} \tag{5.13}$$

$$u_i = \mathbf{c}_{ij}^T \mathbf{x}_{c_j} + d_{ij} e_j \tag{5.14}$$

where $\mathbf{c}_{f_j}^T$ is the jth row of C_f; thus, $\mathbf{c}_{f_j}^T B_f \mathbf{r} = -\lambda_f r_j$ and $\mathbf{c}_{f_j}^T B_w \mathbf{w} = -\lambda_f [1 \ldots 1] \mathbf{w}_j$.

5.3.3.1 Biproper transfer functions $c_{ij}(s)$

If $c_{ij}(s)$ is biproper ($\rho_{ij} = 0$, $d_{ij} \neq 0$), during SM we have

$$e_j = d_{ij}^{-1}(\tilde{u}_i - \mathbf{c}_{ij}^{\mathrm{T}} \mathbf{x}_{c_j}) \tag{5.15}$$

This results from solving (5.14) for e_j and equalling (5.8) to zero. Thus, the last row of (5.13) becomes redundant. Replacing (5.15) in the first r_{d_j} rows of (5.13), the following SMRC dynamics is obtained:

$$\dot{\mathbf{x}}_{c_j} = Q_{c0} \mathbf{x}_{c_j} + \mathbf{b}_j d_{ij}^{-1} \tilde{u}_i$$
$$Q_{c0} = (A_j - \mathbf{b}_j d_{ij}^{-1} \mathbf{c}_{ij}^{\mathrm{T}}) \tag{5.16}$$

where the eigenvalues of Q_{c0} are given by the zeros of $c_{ij}(s)$. Note that this dynamics is hidden ($u_i = \tilde{u}_i$ during SMRC) and it depends only on the controller parameters (not on the plant model). This is because the bounded 'disturbance' y_j (recall $P(s)$ is stable) is completely rejected (see Section 5.3.4).

5.3.3.2 Strictly proper transfer functions $c_{ij}(s)$

To derive the SM dynamics corresponding to strictly proper $c_{ij}(s)$ ($\rho_{ij} \geq 1$, $d_{ij} = 0$), we should express (5.13) and (5.14) in their normal canonical form. Using the transformation studied in Section 2.5.1 yields[2]

$$\begin{cases}
\dot{u}_{i_1} &= u_{i_2} \\
\dot{u}_{i_2} &= u_{i_3} \\
\cdots &= \cdots \\
\dot{u}_{i_{\rho-1}} &= u_{i_\rho} \\
\dot{u}_{i_\rho} &= a_\xi \xi_c + a_\eta \eta_c + b\, e_j \\
\dot{\eta}_c &= P_c \xi_c + Q_c \eta_c \\
\dot{e}_j &= \mathbf{c}_{f_j}^{\mathrm{T}} A_f \mathbf{x}_f + \mathbf{c}_{f_j}^{\mathrm{T}} B_f \mathbf{r} - \dot{y}_j + \mathbf{c}_{f_j}^{\mathrm{T}} B_w \mathbf{w}
\end{cases} \tag{5.17}$$
$$u_i = u_{i_1}$$

where $\xi_c = [u_{i_1}\, u_{i_2} \cdots u_{i_\rho}]^{\mathrm{T}}$ comprises the controller output u_i and its first ($\rho_{ij} - 1$) derivatives, η_c are ($r_{d_j} - \rho_{ij}$) linearly independent states and $b \neq 0$.

When the system operates in its linear region, $\mathbf{w}$ is identically zero and the conditioning loop is inactive. However, when a controller output reaches an

---

[2] Note that ρ_{ij} has been replaced by ρ in the subindexes.

actuator limit, the SMRC activates and the last equation of (5.17) becomes redundant. By making (5.9) equal to zero, the reduced SMRC dynamics results:

$$
\begin{cases}
\dot{u}_{i_1} &= u_{i_2} \\
\dot{u}_{i_2} &= u_{i_3} \\
\cdots &= \cdots \\
\dot{u}_{i_{\rho-1}} &= u_{i_\rho} \\
\dot{u}_{i_\rho} &= \left(\tilde{u}_i - u_i - \sum_{\alpha=1}^{\rho-1} k_\alpha^{ij} u_{i_{\alpha+1}} \right) / k_\rho^{ij} \\
\dot{\eta}_c &= P_c \xi_c + Q_c \eta_c
\end{cases}
\qquad (5.18)
$$
$$
u_i = u_{i_1}
$$

which obviously has the same structure as the SMRC dynamics obtained in (2.46) for strictly proper systems. Recall that in this form, the zeros of $c_{ij}(s)$ are the eigenvalues of Q_c, and so they determine the controller hidden dynamics.

Then, either biproper or strictly proper $c_{ij}(s)$ related to a potential saturating actuator must be MP so that the SMRC dynamics is globally stable. Note, however, that this restriction on individual entries $c_{ij}(s)$ of the controller transfer matrix does not mean that the MIMO controller $C(s)$ or the process $P(s)$ must be MP. In fact, as we have seen in Chapter 4, transmission zeros may be in the RHP while all the zeros of the individual transfer functions are of MP. This is verified by a couple of examples in the next section.

In (5.18), the dynamics of the controller output u_i only depends on the values chosen for k_α^{ij}. If k_α^{ij} are chosen properly, then during SMRC, the controller output u_i will tend towards its saturation limit $\tilde{u}_i$ without overshoots and at a rate determined by k_α^{ij}. Consequently, no differences will exist between **u** and **û** (they will coincide for all time), and the dynamic decoupling of the system will be preserved. Just when the control action falls into the region delimited by the saturation limits without risk of abandoning linear operation, the SMRC loop will become inactive.

5.3.4   Operating issues

5.3.4.1   Disturbance rejection

As we have seen in Section 2.7, a distinctive property of sliding regimes is that they are not affected by disturbances that are collinear with the discontinuous action (i.e. that satisfy the so-called *matching condition*) [114]. It is said that the SM presents *strong invariance* to that kind of disturbances. Observe that this property is associated with the high gain involved in sliding regimes. For our SMRC scheme, r_j and y_j act as matched disturbances for the conditioning loop. This can be seen from the right-hand side of (5.13), where the second and third terms, which can be interpreted, respectively, as the disturbance and the control vector fields of the

SMRC loop, represent collinear vectors. Indeed, using the same notation as that in Section 2.7 for (5.13), we have

$$g(x) = \begin{bmatrix} 0 \\ 1 \end{bmatrix}, \quad u = \mathbf{c}_{f_j}^T B_w \mathbf{w} \tag{5.19}$$

and the two disturbance components are

$$\mu(x) = \mathbf{c}_{f_j}^T A_f \mathbf{x}_f + \mathbf{c}_{f_j}^T B_f \mathbf{r} - \dot{y}_j, \quad \zeta(x) = 0 \tag{5.20}$$

Thus, the SMRC loop for dynamic decoupling preservation satisfies the *matching condition* (2.90).

The above analysis means that, for suitable bounds on $\mathbf{r}$, $\mathbf{x}_f$, $\mathbf{y}$ and $\dot{\mathbf{y}}$, there always exist w_{ij}^+ and w_{ij}^- that ensure condition (2.9), i.e. the SM establishment and maintenance. In this way, r_j and y_j affect neither the dynamics nor the stability of the SMRC loop, as shown by (5.16) and (5.18). This is an important feature of the current approach for stable open-loop plants, since it allows designing the SMRC loop independently of the main loop design.

5.3.4.2 Behaviour of the whole system

During SMRC, the controller output u_i will coincide with or will tend towards the saturation limit $\tilde{u}_i$ with the SMRC dynamics chosen, which is not affected by the main loop because of the robustness properties of the SM mentioned above. Actually, the limit value $\tilde{u}_i$ acts as the input for the conditioning loop, whose output is u_i. Thus, the controlled variable y_j will evolve transiently according to a serial connection of the saturation limit $\tilde{u}_i$ (input), the SMRC dynamics (from $\tilde{u}_i$ to u_i) and the stable dynamics of the plant (from u_i to y_j). Since control directionality does not change and only r_{f_j} is being conditioned, full decoupling is preserved and the other controlled variables remain unaffected. Thus, the whole dynamics will be stable during the transient SM operation.

If the actuators are properly chosen for the control objective, i.e. for every desired reference r_j

$$|[\mathbf{P}(0)]_j \tilde{\mathbf{u}}| > |r_j| \tag{5.21}$$

with $[\mathbf{P}(0)]_j$ being the *j*th row of the DC-gain of the plant, then the available control $\tilde{\mathbf{u}}$ will be sufficient for leading y_j close to its set point r_j.[3] Therefore, as was mentioned, the state trajectory will evolve naturally towards the linear region, and the SMRC loop will become inactive in finite time. From then on, the system recovers the original closed-loop dynamics.

---

[3] Observe that condition (5.21) is equivalent to the sufficient condition (2.39) derived in Chapter 2, since for static decoupling, $S_c(0)^{-1} = Q(0)^{-1}$ must equal $\mathbf{P}(0)$.

5.4   Minimum-phase example

As a first application example, consider again the plant (5.1) given in Example 5.1. As was verified in Figure 5.2, the nominal dynamic decoupling achieved by controller (5.2) for linear operation is completely lost when the control direction changes because of input saturation.

With the objective of evaluating the SMRC method, we add to the system a filter like the one described in (5.10), even for the case where no SM correction is made. This will allow us to compare the system performance with and without SMRC, ruling out the possibility that their differences are due to the added filter.[4]

Differing from the filter used in Example 5.1, the dynamics of this new filter is chosen 10 times faster than the closed-loop one, so that it does not affect the system response during the linear operation of the actuators. The matrices of the state-space representation of the filter are

$$
A_f = -10I_2, \quad C_f = 10I_2
$$

$$
B_f = I_2, \quad B_w = \begin{bmatrix} 1 & 1 & 0 & 0 \\ 0 & 0 & 1 & 1 \end{bmatrix}
\tag{5.22}
$$

where B_w only makes sense for SMRC compensation. We also make input constraints harder, taking $\tilde{u}_1 = \tilde{u}_2 = \pm 12$.

The unconstrained case will then achieve a faster closed-loop response than the one shown in Figure 5.1, whereas the input saturation will cause a much greater degradation than what is shown in Figure 5.2 because of the higher filter cut-off frequency and the tighter constraints. However, since the conclusions are the same as those in Example 5.1, these responses are not shown here.

To solve the control directionality problem, we apply the SMRC loop proposed in this chapter to controller (5.2). In this case, as both actuators have the same limits ($\bar{u}$ and $\underline{u}$ can be taken as scalars) and the maximum relative degree of the controller transfer matrix is zero ($\rho_{ij} = 0$, $i, j = 1, 2$), the constant coefficient matrix M is reduced to

$$
M = \begin{bmatrix}
-1 & 0 & 0 & 1 \\
0 & -1 & 0 & 1 \\
-1 & 0 & 0 & 1 \\
0 & -1 & 0 & 1 \\
-1 & 0 & 1 & 0 \\
0 & -1 & 1 & 0 \\
-1 & 0 & 1 & 0 \\
0 & -1 & 1 & 0
\end{bmatrix}
\tag{5.23}
$$

---

[4] A filter that avoids, by its own, the input saturation would result in an extremely conservative controller for small set-point changes.

whose input is $[u_1 \ u_2 \ \underline{u} \ \bar{u}]^{\mathrm{T}}$. Observe that the first four rows of M give rise to surfaces $\overline{\sigma}_{ij}$, whereas the remaining ones generate σ_{ij}. Each component of the discontinuous action $\mathbf{w} = [w_{11} \ w_{21} \ w_{12} \ w_{22}]^{\mathrm{T}}$ commutes according to the switching law (5.7) between $w_{ij}^- = -1$, zero and $w_{ij}^+ = 1$.

The effectiveness of the SMRC compensation is exhibited in Figure 5.6. The system outputs remain dynamically decoupled (solid lines) and with a performance that is close to the performance of the ideal unconstrained case (dashed lines for comparative purposes). Likewise, controller outputs never surpass the actuator physical limits, coinciding at each time instant with the plant inputs (u and $\hat{u}$ are overlapped in the lower plot of the figure).

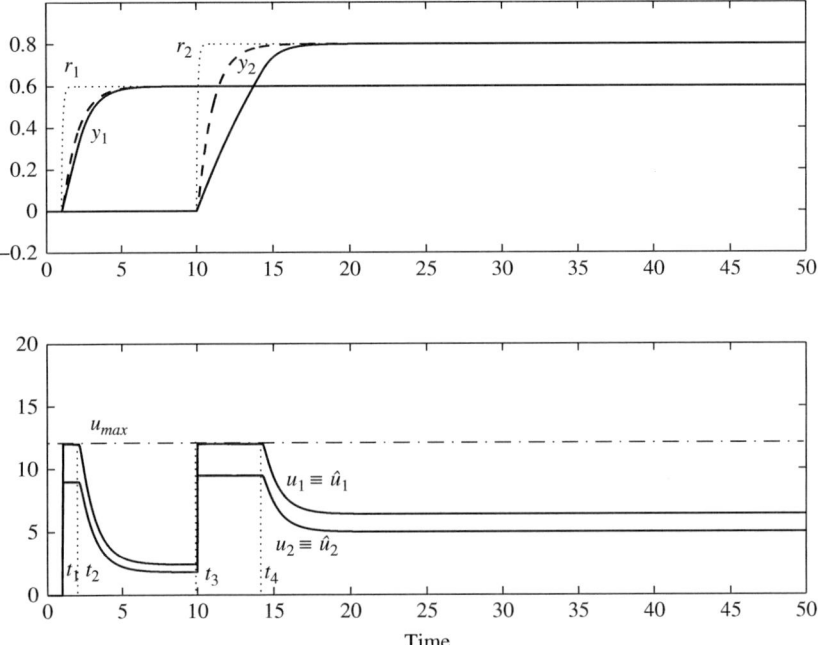

Figure 5.6   Evolution of the controlled variables y and the control actions u for the constrained system with SMRC

When reference r_1 changes and u_1 reaches the upper limit of the first actuator, SMRC activates over the surface $\overline{\sigma}_{11} = 0$ between t_1 and t_2, which avoids u_1 saturation by the proper conditioning of r_1. Afterwards, the SMRC loop acts again after the change in r_2, now over the surface $\overline{\sigma}_{12} = 0$ from t_3 to t_4. Note that the switching law ((5.7) and (5.8)) avoids conditioning the invariant reference in each case, and that the attenuation of u_2 is not because of SM on surfaces $\overline{\sigma}_{2j} = 0$, but a consequence of the conditioning performed first to r_1 and then to r_2 to avoid u_1 saturation.

5.5   Non-minimum phase examples

5.5.1   *Revisiting Example 1.3*

Consider once more the plant used to introduce the control directionality problem
in Chapter 1. The transfer matrix of the plant was

$$P(s) = \frac{1}{(s+1)^2} \begin{bmatrix} s+2 & -3 \\ -2 & 1 \end{bmatrix} \tag{5.24}$$

At this part of the book, we can easily verify that $P(s)$ has a transmission zero
at $s = 4$. Thus, the ideas presented in Section 4.3.2 can be followed to
synthesise controller (1.9), which achieves the decoupled complementary sensitivity (1.8).

We aim now at improving the response obtained in Figure 1.5 in the presence
of input saturation. A fast prefiltering of the reference signal, also present in
Example 1.3, was included. It can be represented by the state-space matrices:

$$A_f = -20 \cdot I_2, \quad C_f = 20 \cdot I_2,$$
$$B_f = I_2, \quad B_w = \begin{bmatrix} 1 & 1 & 0 & 0 \\ 0 & 0 & 1 & 1 \end{bmatrix} \tag{5.25}$$

The same negative change as that in Figure 1.5 is applied to reference r_2, but now
with the SMRC algorithm. The matrix M (without considering the transformation
needed to obtain the derivatives from the controller states) is given by

$$M = \begin{bmatrix} -1 & 0 & 0 & 0 & 1 & 0 \\ 0 & -1 & 0 & 0 & 1 & 0 \\ -1 & 0 & -0.02 & 0 & 1 & 0 \\ 0 & -1 & 0 & 0 & 1 & 0 \\ -1 & 0 & 0 & 0 & 0 & 1 \\ 0 & -1 & 0 & 0 & 0 & 1 \\ -1 & 0 & -0.02 & 0 & 0 & 1 \\ 0 & -1 & 0 & 0 & 0 & 1 \end{bmatrix} \tag{5.26}$$

where the input vector is $[u_1 \ u_2 \ \dot{u}_1 \ \dot{u}_2 \ \underline{u} \ \bar{u}]^T$, with $\tilde{u} = \tilde{u}_i = \pm 2$, $i = 1, 2$. We take
here $w_{ij}^- = 1$ and $w_{ij}^+ = -1$ (these 'inverted' values are related to the negative gains
in the transfer functions of $C(s)$). Observe that in this case, only $\underline{\sigma}_{12}$ and $\bar{\sigma}_{12}$ include
$\dot{u}_1$, in concordance with the relative degree structure of controller (1.9).

The left side of Figure 5.7 exhibits the achievements of SMRC loop. Dynamic
decoupling is preserved even in the presence of constraints, and furthermore, y_2
response is only gracefully affected (dashed lines correspond to the ideal unconstrained case). Recall that the inverse response is the price to be paid for dynamic
decoupling of NMP systems, as seen in Chapter 4. It is important to remark that the
controller outputs do never exceed the actuator limits, and therefore they coincide
at every instant of time with the plant inputs. After the step in r_2, SMRC activates

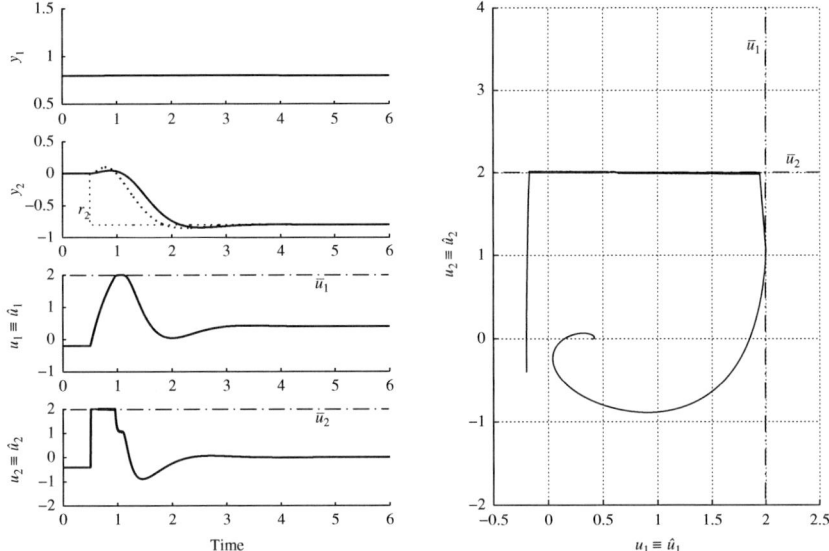

*Figure 5.7   Response of decoupled NMP system with input saturation and SMRC.
See Figure 1.5 for original response*

first on the surface $\bar{\sigma}_{22} = 0$ (limiting u_2) and then on $\bar{\sigma}_{12} = 0$ (bounding u_1). The constant $k_1^{12} = -0.02$ in the latter surface is chosen for a time constant of approximately 20 ms (hardly noticeable in the figure).

The right plot in Figure 5.7 evidences the control directionality preservation. Unlike the case given in Figure 1.5, $\hat{u} \equiv u$ when SMRC is applied. Note that when approaching $\bar{u}_1$ from $u_2 = 2$, the SMRC starts before the limit is reached because of the dynamics imposed by $\bar{\sigma}_{12} = 0$, which manages the approaching speed to the constraint.

Finally, Figure 5.8 shows the reference r_{f2} with SMRC (solid line) and without SMRC (dotted line), and the activated switching signals w_{12} and w_{22} corresponding to Figure 5.7.

5.5.2   Sugar cane crushing station

As a last example, we present an industrial application of the proposed SMRC strategy. In this case, the methodology is applied to the sugar cane milling process. Figure 5.9 presents a schematic diagram of the crushing stages that can be found in many sugar factories.

For maximal juice extraction, the buffer chute height $h(t)$ and the mill torque $\tau(t)$ are controlled by means of the turbine speed $\Omega(t)$ and the position $f(t)$ of a hydraulic flap mechanism, which adjusts the chute volume. Although the control of the torque has significant influence on juice extraction, the main purpose of the chute height regulation is to filter out the main disturbance $d_0(t)$, generated by the fluctuating feed of sugar cane to the buffer chute. Actually, the chute height should

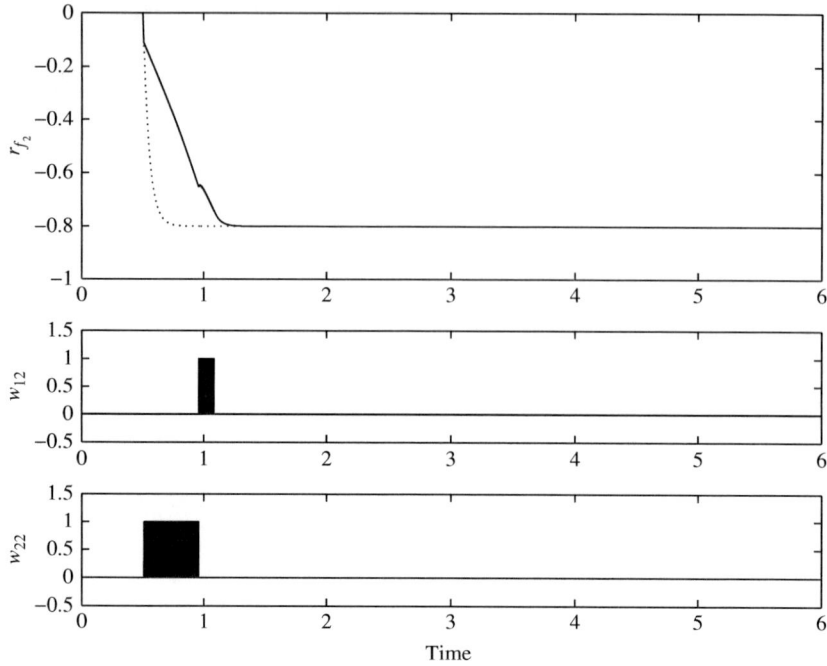

Figure 5.8   Conditioned reference r_{f2} and discontinuous signals w_{i2} corresponding to Figure 5.7

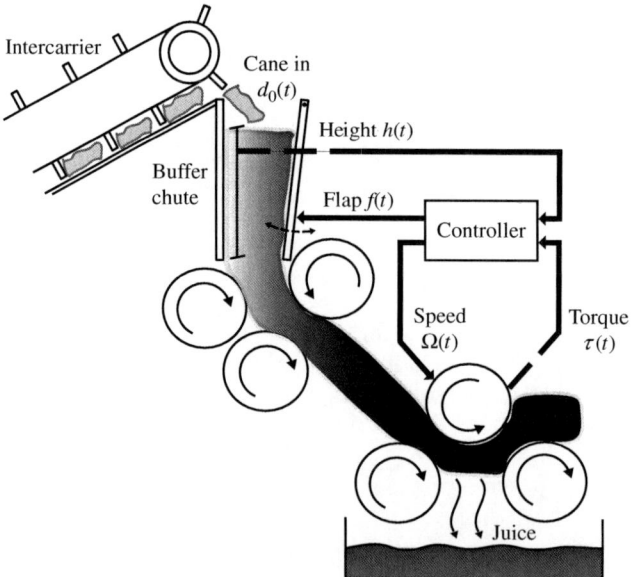

Figure 5.9   Sugar mill diagram

be kept within a given range, where the cane compaction is adequate for an effi-
cient milling, rather than being regulated [93].

The following linearised model was obtained for an Australian crushing station
as the one given in Figure 5.9 from experimental data [148]:

$$\begin{bmatrix} \tau(t) \\ h(t) \end{bmatrix} = \begin{bmatrix} -\frac{5}{25s+1} & \frac{s^2-0.005s-0.005}{s(s+1)} \\ \frac{1}{25s+1} & -\frac{0.0023}{s} \end{bmatrix} \begin{bmatrix} f(t) \\ \Omega(t) \end{bmatrix} + \begin{bmatrix} -\frac{0.005}{s} \\ -\frac{0.0023}{s} \end{bmatrix} d_0(t)$$

This model has an RHP zero at $s = 0.137$ with associated direction $h^T = [1\ 5]^T$.
The strong alignment of the zero with the 'secondary' variable $h(t)$ indicates that
triangular decoupling is suitable for this process, since small interactions will occur
in $h(t)$ if only the 'main' variable $\tau(t)$ is decoupled. Furthermore, such a design will
also push the NMP effect to the second variable $h(t)$, avoiding the spreading over
$\tau(t)$ that results from full dynamic decoupling.[5] This was verified in Reference 49,
where among the different designs tested on this plant, triangular decoupling achieved
achieved the best performance.

We then took from Reference 49 the controller that attained triangular
decoupling and ran simulations with it. The response of the closed-loop system
with ideal unconstrained actuators is shown in Figure 5.10. It can be observed how
the torque $\tau(t)$ is unaffected by changes in the chute height reference r_{f_h}. Furthermore, the RHP zero spreading is avoided: only the chute height evolution shows
step inverse response. The closed loop also compensates step disturbance $d_0(t)$
rapidly, particularly in the torque channel. However, the bottom half of the figure
shows that the turbine speed $\Omega(t)$ presents quite large and sudden changes after the
step in the height channel, giving rise to a potential input saturation. The greater the
bandwidth demands, the higher the risk of saturation.

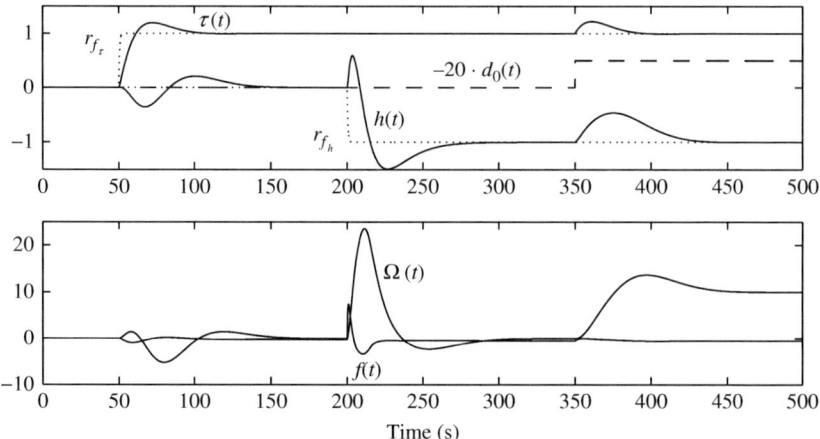

Figure 5.10   Partially decoupled response with ideal unconstrained actuators

---

[5] For further details on these features of triangular decoupling designs, see Chapter 7.

Then, we introduce an isolated saturation element to the turbine speed to evaluate its effects on system performance. As Figure 5.11 shows, just a slight saturation of the turbine speed $\Omega(t)$ leads to great interactions with the torque when

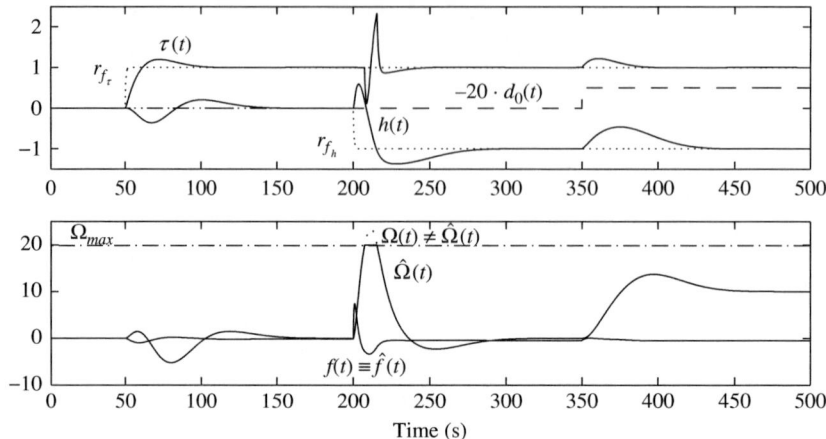

Figure 5.11   Response degradation due to turbine speed saturation

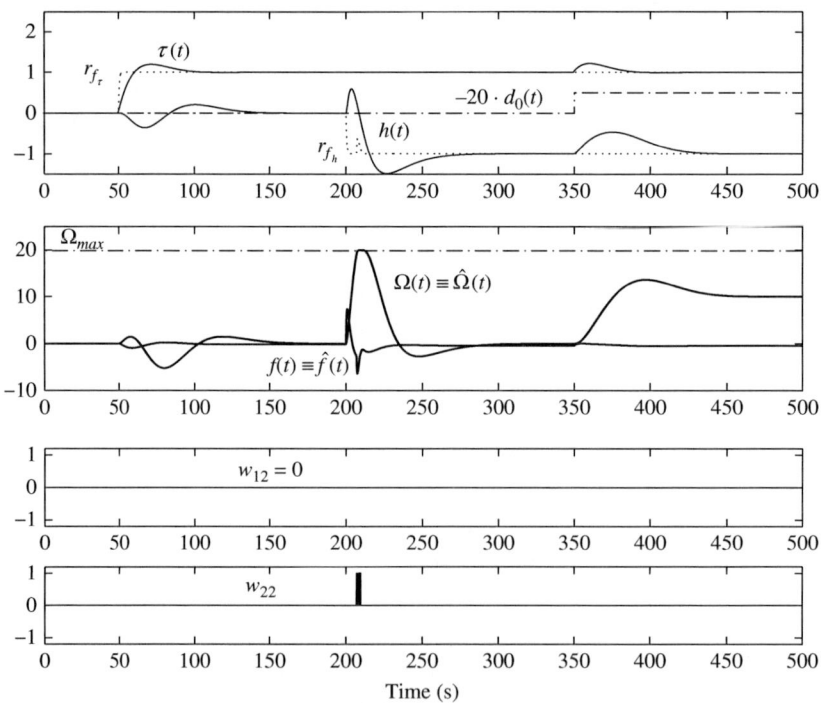

Figure 5.12   Improvement of the constrained system response by means of SMRC approach

a step is applied to the height set point. Like in the previous examples, the desired decoupling is lost as a consequence of input saturation and the associated change of control directionality.

The proposed SMRC compensation is then added, this time in a simplified configuration. Because only torque decoupling preservation is desired, it is sufficient to generate discontinuous signals w_{12} and w_{22} to shape $r_{f_h} = r_2$ when $f(t) = u_1$ or $\Omega(t) = u_2$ reach their bounds, respectively. The results presented in Figure 5.12 reveal that the SMRC approach effectively preserves triangular decoupling in the presence of turbine speed constraints. To this end, the SMRC loop briefly shapes r_{f_h} with the discontinuous signal w_{22}. Note that disturbance rejection is not affected at all since the original closed-loop performance is recovered once the system re-enters the linear region.

The reader interested in sugar cane factories can also find in Reference 43 an application of SMRC ideas to improve performance in older sugar mills. Therein, the only manipulable input is the belt conveyor speed, and the objective is to delimit chute height (output) variations when facing discontinuous sugar supply.

Chapter 6

Interaction limits in decentralised control architectures

The methods for reducing or cancelling crossed interactions studied so far in multivariable systems were based on centralised multiple-input multiple-output (MIMO) controllers, which is also usual in multivariable control literature. However, despite the performance advantages of centralised controllers, the great majority of industrial process control applications still rely on decentralised or multiloop control structures. Because of their *structural constraints*, decentralised controllers are not able to suppress by their own interactions of the plant, which are only taken into account at the controller tuning phase. This is not a trivial problem to be solved. In fact, even when supervisory control tools like model predictive control are employed, the coupling among the loops has to be addressed at the lower level (generally decentralised PI/PID control) because of the long sampling time of the supervisory modes. Therefore, the coupling reduction under decentralised structures is a topic of great interest when considering practical control issues. In this chapter, we shall analyse and address this problem.

First, some basic concepts related with this control topology are presented. In particular, the relative gain array (RGA) is introduced as a simple interaction measure, whereas the potential effects of crossed interactions are illustrated through a simple example. Then, the sliding mode reference conditioning (SMRC) technique is exploited here in order to limit the amplitude of decentralised control interactions.

6.1 Introduction to decentralised control

6.1.1 Architecture description

Let $P(s) \in \mathbb{R}^{m \times m}(s)$ relate the plant input vector $u(t) = [u_1 \, u_2 \, \ldots \, u_m]^{\mathrm{T}}$ with the output vector $y(t) = [y_1 \, y_2 \, \ldots \, y_m]^{\mathrm{T}}$, and let $r(t) = [r_1 \, r_2 \, \ldots \, r_m]^{\mathrm{T}}$ be the vector of reference signals or *set points* for the MIMO control system. Then, it is said that the control of $P(s)$ is decentralised if the controller $C(s)$ can be represented by a diagonal transfer matrix, i.e. $C(s) = diag(c_1(s), c_2(s), \ldots, c_m(s))$, and

$$u_i(s) = c_i(s)(r_i(s) - y_i(s)) \tag{6.1}$$

The general scheme given in Figure 6.1 describes a decentralised control system, also known as multiloop control. Naturally, the structural constraints

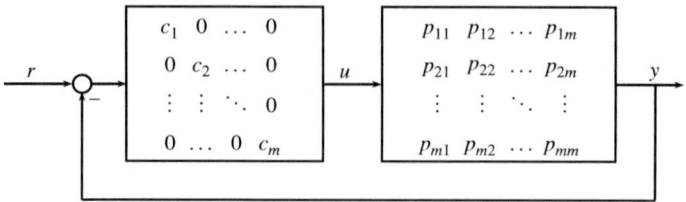

$$\begin{bmatrix} c_1 & 0 & \cdots & 0 \\ 0 & c_2 & \cdots & 0 \\ \vdots & \vdots & \ddots & 0 \\ 0 & \cdots & 0 & c_m \end{bmatrix}$$

$$\begin{bmatrix} P_{11} & P_{12} & \cdots & P_{1m} \\ P_{21} & P_{22} & \cdots & P_{2m} \\ \vdots & \vdots & \ddots & \vdots \\ P_{m1} & P_{m2} & \cdots & P_{mm} \end{bmatrix}$$

Figure 6.1   General structure of a decentralised control system

imposed to the controller lead to a closed-loop performance degradation with respect to centralised multivariable controllers. Even then, most multivariable control loops in industry are based on decentralised architectures because of their practical advantages: flexibility in operation, failure tolerance, simplified design and tuning, etc. [17]. Also, communication nets, startup schemes and identification experiments are considerably harder to face with centralised control than with decentralised control.

On the other hand, given its great popularity in practical applications, some engineers and control practitioners tend to think that multiloop strategies are enough for controlling any kind of MIMO system. In the next section, we will illustrate by means of a simple numerical example that although most *real-life* systems use this architecture, the decentralised control does not always lead to satisfactory solutions. In some cases the inherent limitations of decentralised control can be overcome by complementary strategies that improve the diagonal controller performance, as the one presented later in this chapter. However, for many other cases there will be no alternative but to consider a centralised controller.

In the following sections, we will study some of the basic properties of decentralised control systems for a better understanding of their advantages and limitations.

6.1.2   Interaction measure

One of the initial problems to be addressed when employing a decentralised control architecture is the 'pairing' of variables, i.e. to decide which input variables (control actions) will be used to control each of the process outputs (controlled variables). In the case of multivariable plants represented by $P(s) \in \mathbb{R}^{m \times m}(s)$, there will be $m!$ possible combinations of inputs and outputs.

Useful tools to solve the pairing problem are the so-called *interaction measures*. One of the first interaction measures, widely spread and employed during long time in practical applications, is the already-mentioned RGA. A useful summary of the main closed-loop properties that can be extracted from the RGA can be found in Reference 70 and references therein.

Here, we follow the reasoning given by Bristol in Reference 12 where he proposed the RGA. It is important to remark that in the RGA development, Bristol considered processes described by constant gain matrices, and therefore the matrix of steady-state gains is generally used to compute the RGA. Indeed, the main idea behind the RGA is to quantify how much the steady-state gain of a given loop is affected by the remaining loops. Nevertheless, posterior works have shown the importance that the RGA has for frequency-dependent analysis [86].

Let u_j and y_i be given input and output, respectively, of a multivariable plant with steady-state gain matrix $P(0)$. Assume that we want to use u_j to control y_i. Since we are dealing with a decentralised structure, following two extreme cases arise:

1. That the other loops are open. Then, all other inputs can be considered constant. In particular, $u_k = 0$, $\forall k \neq j$.
2. That the other loops are closed. In this case, assuming perfect control in these loops, all the remaining outputs can be considered constant. Particularly, $y_k = 0$, $\forall k \neq i$.

Note that although the perfect control ($y_k = r_k$, r_k reference of the kth loop) is only achievable in steady state, it can be considered as a good approximation for frequencies below the bandwidth of each loop.

We now evaluate the effect $\partial y_i / \partial u_j$ of input u_j on output y_i for the cases 1 and 2. Hence, we have

1. With the other loops open, it is straightforward from $y = P(0)u$ that

$$\left(\frac{\partial y_i}{\partial u_j}\right)_{u_k=0;k\neq j} = p_{ij}(0) \tag{6.2}$$

That is, the effect of u_j on y_i is given by the element ij of the matrix $P(0)$, which is denoted by $p_{ij}(0) = [P(0)]_{ij}$.

2. In order to evaluate this effect when the other loops are closed, we commute the roles of $P(0)$ and $P^{-1}(0)$, of u and y, and of i and j. In this way, starting now from $u = P^{-1}(0)y$ we have

$$\left(\frac{\partial u_j}{\partial y_i}\right)_{y_k=0;k\neq i} = [P^{-1}(0)]_{ji} \tag{6.3}$$

resulting finally for this case

$$\left(\frac{\partial y_i}{\partial u_j}\right)_{y_k=0;k\neq i} = 1/[P^{-1}(0)]_{ji} = \hat{p}_{ij}(0) \tag{6.4}$$

The RGA is then defined as the matrix whose elements are the relative gains

$$\lambda_{ij} \triangleq \frac{p_{ij}(0)}{\hat{p}_{ij}(0)} = [P(0)]_{ij}[P^{-1}(0)]_{ji} \tag{6.5}$$

Denoting the RGA as Λ yields

$$\Lambda = P(0) \times [P^{-1}(0)]^{\mathrm{T}} \tag{6.6}$$

where $\times$ is the Schur product (element-by-element multiplication).

As can be appreciated from the above definition, it will be convenient that the relative gain λ_{ij} between the variable to be controlled y_i and the input chosen for its control u_j is close to 1, since this means that the gain from u_j to y_i is almost unaffected by closing the other loops. On the other hand, a gain $\lambda_{ij} < 0$ indicates that the steady-state gain between u_j and y_i will change its sign when the other loops are closed. Thus, the pairing of variables with relative gain $\lambda_{ij} < 0$ should be avoided.

Remark 6.1: *From the trivial definition in (6.6), $\Lambda(s) = P(s) \times [P^{-1}(s)]^\mathrm{T}$ has then also been employed as an interaction measure with frequency information. This one, and other frequency-dependent interaction measures can be found, for instance, in References 19, 61, 75, 86 and 116.*

The RGA has, among others, the following important algebraic properties:

1. Its row and column elements sum to one.
2. Commutation of columns (rows) in $P(s)$ leads to the same commutation of columns (rows) in $\Lambda(s)$.
3. The RGA is independent of input and output scaling (the scaling of a matrix P is performed by multiplying the matrix by two non-singular diagonal matrices D and D' so that the scaled matrix results in $P' = DPD'$).
4. A change in an element of $P(s)$ equal to the negative inverse of its corresponding RGA element yields singularity.
5. The RGA is equal to the identity matrix if $P(s)$ is either upper or lower triangular.

The last property shows that $\Lambda(s) - I$ gives a measure of *two-way interactions*, i.e. those cases in which the loops interact with each other. This does not occur in a plant with triangular transfer matrix, where each loop affects only those loops that have no effect on it (*one-way interaction*).

6.1.3   Control structure selection: the TITO case

Now consider the simplified case where the system has two inputs and two outputs. These systems are known as TITO (*two inputs–two outputs*) systems. Figure 6.2 represents a block diagram of a TITO system with a multiloop control structure.

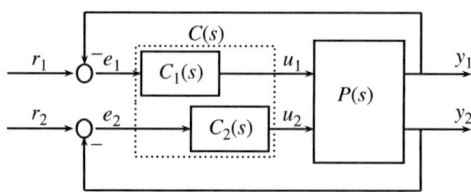

Figure 6.2   TITO system under decentralised control structure

For these systems, the first RGA property listed in Section 6.1.2 implies that the RGA is determined by a single scalar parameter λ, known as *Bristol interaction index*. A TITO system can be represented by the following equations:

$$y_1(s) = p_{11}(s)u_1(s) + p_{12}(s)u_2(s)$$
$$y_2(s) = p_{21}(s)u_1(s) + p_{22}(s)u_2(s)$$

(6.7)

or by means of the transfer matrix

$$P(s) = \begin{bmatrix} p_{11}(s) & p_{12}(s) \\ p_{21}(s) & p_{22}(s) \end{bmatrix}$$

(6.8)

Let us see how the static gain of the first loop is affected by the second loop. According to the Bristol assumption of having perfect control when closing the second loop, so that gains of the extreme cases are non-dependent of the type of controller, one may write

$$y_1(s) = p_{11}(s)u_1(s) + p_{12}(s)u_2(s)$$
$$0 = p_{21}(s)u_1(s) + p_{22}(s)u_2(s)$$

(6.9)

By eliminating $u_2(s)$ from the first equation, the gain of the first loop when the second loop is closed yields

$$y_1(s) = \frac{p_{11}(s)p_{22}(s) - p_{12}(s)p_{21}(s)}{p_{22}(s)} u_1(s)$$

(6.10)

The ratio between the static gain of the first loop with the second loop open ($p_{11}(0)$) and with the second loop closed

$$\lambda = \frac{p_{11}(0)p_{22}(0)}{p_{11}(0)p_{22}(0) - p_{12}(0)p_{21}(0)}$$

(6.11)

is the *Bristol interaction index*. Despite referring to static conditions, this parameter can be used as a guide in order to decide how the controllers should be connected in a loop-by-loop TITO control system, and it is particularly meaningful for low-frequency signals. If $\lambda = 1$, which corresponds to $p_{12}(0)p_{21}(0) = 0$, there is no interaction (the static gain of the first loop is not affected by the second loop).

Evidently, the RGA for the system (6.7) can be written in terms of the Bristol interaction index as

$$\Lambda = \begin{bmatrix} \lambda & 1-\lambda \\ 1-\lambda & \lambda \end{bmatrix} \tag{6.12}$$

with λ given by (6.11).

If $0 < \lambda < 1$, the gain of each loop is greater when the other loop is closed than when the other loop is open. However, if $\lambda > 1$, the loop gains decrease when the other loop is closed. Again, the worst case is the condition $\lambda < 0$, which means that the static gain of a given loop changes its sign when the other loop is closed.

Bristol showed that the best controller connection is the one that makes the relative gains positive and as close to one as possible. Note that if $\lambda = 0$, there is no interaction, but the loops should be interchanged. In general, the input/output connection should be interchanged when $\lambda < 0.5$ [4]. In Reference 86, McAvoy postulated that a pairing with $0.67 < \lambda < 1.5$ should be achieved for the closed-loop performance to be acceptable. More recently, Åström and Hägglund suggested decoupling the system if the Bristol index is outside this interval, since a decentralised control will not lead to satisfactory results [4]. In general, the plants with large RGA elements are hard to control [118], especially by means of decentralised controllers. Although decoupling can still be used [45], the task is difficult even for centralised controllers.

Remark 6.2: *There exists a strong correlation between the RGA (or a norm of the RGA) and the condition number $\kappa(P)$, which is defined as the ratio between the maximum and minimum singular values of a multivariable system. When a system has a large $\kappa(P)$, it is said that the system is ill-conditioned. The relations between these two indexes were stated in References 52 and 92, showing that a system with high RGA elements is always ill-conditioned. This undesired system trait must then be avoided as far as possible by an appropriate process design and a suitable actuator selection [88].*

Example 6.1: Let us consider again the systems given in Example 4.2

$$P_1(s) = \frac{1}{(1+s)^2} \begin{bmatrix} s+3 & 2 \\ 3 & 1 \end{bmatrix} \tag{6.13}$$

and Example 4.3

$$P_2(s) = \frac{1}{(s+1)(s+2)} \begin{bmatrix} s-1 & s \\ -6 & s-2 \end{bmatrix} \tag{6.14}$$

The RGA of the non-minimum phase (NMP) multivariable system $P_1(s)$ is

$$\Lambda = \begin{bmatrix} -1 & 2 \\ 2 & -1 \end{bmatrix} \tag{6.15}$$

and thus its Bristol index $\lambda = -1$. According to the above discussion, the system outputs should be interchanged to control the system loop by loop, although the diagonal elements do not present a priori any difficulty as individual transfer functions.

On the other hand, the RGA of $P_2(s)$ is the identity matrix and its Bristol interaction index $\lambda = 1$, something that may seem counter-intuitive considering the NMP zeros of the system diagonal elements. This indicates that in a multiloop structure, y_1 should be controlled with u_1 and y_2 with u_2.

It is also interesting to note that, as was seen in Example 4.3, the system $P_2(s)$ has the multivariable zeros at $s = -1$ and $s = -2$. We have then an interesting case in which the interactions can have a beneficial effect: while the individual control of both the first and the second loop presents severe limitations due to the right-half plane (RHP) zeros in $p_{11}(s)$ and $p_{22}(s)$, the centralised multivariable control may achieve faster responses since its zeros are in the left-half plane (LHP). However, one should be careful when exploiting this interaction 'help', since in case a loop is broken for some reason the closed-loop system may become unstable, as the remaining single NMP loop would be left with an excessively large controller gain. This situation is further analysed in Reference 50, where time-domain constraints of decentralised control systems useful for computing settling-time and inverse-response bounds are also derived.

6.1.4 Decentralised integral controllability

Another desirable feature of decentralised control systems is that a given loop can be retuned or disconnected without the closed-loop system becoming unstable. This is very important in practice because it allows obtaining the desired closed-loop performance by means of a loop-by-loop tuning without the risk of the system becoming unstable. Besides, it prevents closed-loop system instability as a consequence of input saturation or sensor/actuator failures.

This property can be better specified by defining the decentralised integral controllability [17,117].

Definition 6.1 (Decentralised integral controllable (DIC) systems): *A plant $P(s)$ is DIC if there exists a decentralised controller with integral action in each loop, such that (a) the feedback system is stable and (b) the gains of each individual loop can be affected by a factor ε_i, $0 < \varepsilon_i < 1$, without leading to closed-loop instability.*

Although there are no necessary and sufficient conditions for an $n \times n$ plant to be DIC, the RGA provides a useful tool to verify the cases in which a plant with a particular control structure is not DIC. This is established by Theorems 6.1 and 6.2, originally proved in Reference 52.

Theorem 6.1: *Consider a stable square plant P(s) and a diagonal controller C(s) with integral action in each loop. Then, if a diagonal RGA element λ_{ii} is negative, the closed-loop system satisfies at least one of the following properties:*

(A)   *The closed-loop system is unstable.*
(B)   *The ith loop is unstable by itself (with the other loops open).*
(C)   *The closed-loop system is unstable if the ith loop is open.*

Clearly, none of the three alternatives that result from a $\lambda_{ii} < 0$ is desired. The worst case is A, but the case C is also critical since it implies that the system becomes unstable if the ith loop is open by some reason, as could be the activation of an input saturation. Therefore, Theorem 6.1 gives another argument to avoid λ_{ii} being lower than zero.

According to Theorem 6.1, inputs and outputs should be connected so that all the relative gains are positive in order to have a DIC system. Note that when evaluating different pairing alternatives, it is not necessary to recompute RGA for each possible pairing choice, since any permutation of columns or rows in $P(s)$ gives rise to the same permutation of columns or rows in the RGA (due to the second of the listed algebraic properties). This allows eliminating most of the possible input/output combinations by simple inspection of the RGA, as shown in Example 6.2.

Example 6.2: Consider a 3 × 3 plant with the following steady-state gain matrix

$$P(0) = \begin{bmatrix} 4.3 & 10.6 & 2.4 \\ 14.1 & -9.6 & -1.2 \\ 8.0 & 1.4 & 2.7 \end{bmatrix} \tag{6.16}$$

Its RGA is

$$\Lambda = \begin{bmatrix} 0.28 & \mathbf{1.34} & -0.61 \\ \mathbf{0.94} & -0.19 & 0.25 \\ -0.22 & -0.14 & \mathbf{1.36} \end{bmatrix} \tag{6.17}$$

For a 3 × 3 plant, there exist six possible ways of connecting a decentralised controller. However, from the RGA we can see that there is only one positive element in the second column ($\lambda_{12} = 1.34$) and in the third row ($\lambda_{33} = 1.36$), thus we have only one connection leading to positive diagonal RGA elements (u_1 with y_2, u_2 with y_1 and u_3 with y_3). In this manner, by a simple glance at the RGA we can eliminate five of the six possible pairings.

For the particular case of TITO system we certainly have a necessary and sufficient condition for a plant $P(s)$ to be DIC [17,117].

Theorem 6.2: *A 2 × 2 plant P(s) is DIC if and only if the Bristol interaction index λ is strictly greater than zero ($\lambda > 0$).*

This result strengthens the importance of performing an appropriate selection of the control pairing in multiloop approaches.

6.2   Interaction effects on multiloop strategies

In this section, we present an illustrative textbook example to show some of the effects that crossed interactions may have on a decentralised control system.

Consider the following benchmark system that has been extensively used in the literature to explain multivariable control issues and is known as *Rosenbrock system* [106]:

$$P(s) = \begin{bmatrix} p_{11}(s) & p_{12}(s) \\ p_{21}(s) & p_{22}(s) \end{bmatrix} = \begin{bmatrix} \dfrac{1}{s+1} & \dfrac{2}{s+3} \\ \dfrac{1}{s+1} & \dfrac{1}{s+1} \end{bmatrix} \tag{6.18}$$

As can be appreciated, the SISO subsystems of this 2×2 process do not present a priori any difficulty in order to be controlled when they are analysed separately. For example, a PI controller will be enough to achieve sufficiently fast responses on y_1 if the second loop is open. However, the MIMO system has an RHP multivariable zero at $s = z_0 = 1$ with output direction $h^T = [1 \ 1]$, which imposes closed-loop performance limitations for the overall MIMO system, particularly for bandwidths greater than $\omega_0 = 1$.

In Reference 4, Åström and Hägglund evaluate the performance achieved by a multiloop PI controller with proportional gain $K_p = 2$ and integral gain $K_i = 2$ (in each loop) on the Rosenbrock system (6.18). The transfer matrix of such a decentralised controller is, obviously

$$C(s) = \begin{bmatrix} \dfrac{2s+2}{s} & 0 \\ 0 & \dfrac{2s+2}{s} \end{bmatrix} \tag{6.19}$$

We propose here a modified version of the plant (6.18). The objective is to gain flexibility to evaluate the interaction effects when the plant is controlled by a diagonal controller like (6.19). With this aim, we propose that the crossed transfer functions $p_{12}(s)$ and $p_{21}(s)$ in (6.18) depend on constant gains k_{12} and k_{21}, which can in fact be interpreted as the different gains that result in practice from linearising the model of a process around different operating points. The proposed plant is then

$$P(s) = \begin{bmatrix} \dfrac{1}{s+1} & \dfrac{k_{12}}{s+3} \\ \dfrac{k_{21}}{s+1} & \dfrac{1}{s+1} \end{bmatrix} \tag{6.20}$$

The multivariable zeros of the system can be obtained from (see Section 4.1.2 in Chapter 4)

$$det(P(s)) = \dfrac{1}{(s+1)^2} - \dfrac{k_{21}k_{12}}{(s+1)(s+3)} = \dfrac{(1-k_{21}k_{12})s+3-k_{21}k_{12}}{(s+1)^2(s+3)} = 0 \tag{6.21}$$

Hence, the system will have a transmission zero at

$$z_0 = s = \dfrac{-3 + k_{21}k_{12}}{1 - k_{21}k_{12}} \tag{6.22}$$

which verifies the presence of the zero at $s = 1$ for the particular case of the Rosenbrock system ($k_{12} = 2$, $k_{21} = 1$). From (6.22), the system (6.20) will be NMP as long as $1 < k_{12}k_{21} < 3$. Besides, the RGA will be

$$\Lambda = P(0) \times [P^{-1}(0)]^{\mathsf{T}} = \frac{1}{3 - k_{12}k_{21}} \begin{bmatrix} 3 & -k_{12}k_{21} \\ -k_{12}k_{21} & 3 \end{bmatrix} \quad (6.23)$$

Thus, the Bristol interaction index is given by $\lambda = 3/(3 - k_{12}k_{21})$. Note that λ changes its sign for $k_{12}k_{21} > 3$.

We consider now the control of this plant by means of the controller (6.19) for five different 'operating points' corresponding to the following sets of gains $k_{12} - k_{21}$:

1. $k_{12} = 0$, $k_{21} = 0$: This is the trivial case in which the MIMO plant is composed of two SISO systems completely decoupled. Then, no interaction will be seen between the loops, and naturally $\lambda = 1$. The system has no zeros, and there are no performance limitations for its control. Assuming unbounded control actions, the closed loop could be made as fast as needed by increasing the PI controller gains. The simulations corresponding to this case are shown in dashed lines in Figure 6.3. The first loop was excited with a unitary step at $t = 1$, while another step reference was applied to the second loop at $t = 15$ (reference signals are plotted with dotted lines).

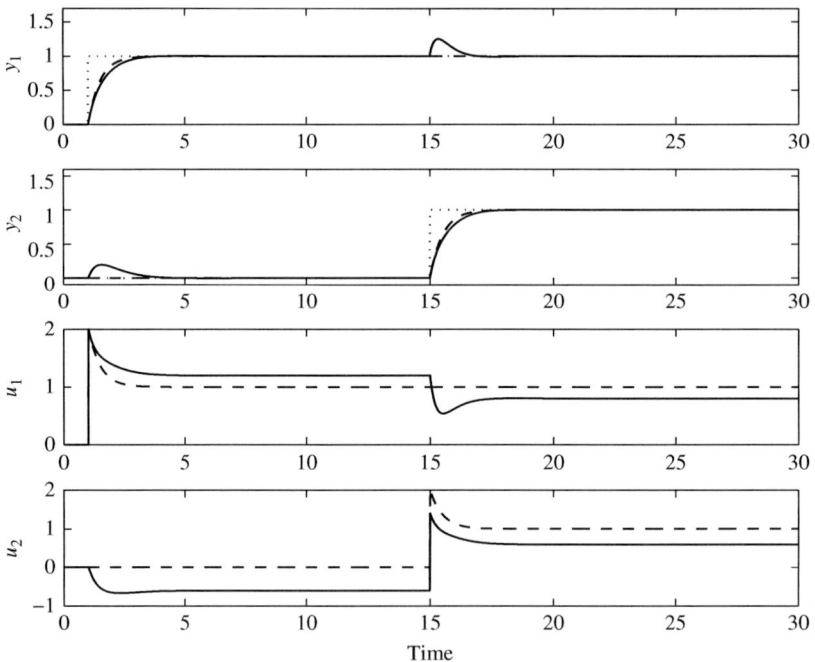

Figure 6.3   Closed-loop response of decentralised PI controller (6.19) and process (6.20) with $k_{12} = 0$ and $k_{21} = 0$ (dashed line) and $k_{12} = 1$ and $k_{21} = 0.5$ (solid line)

2. $k_{12} = 1$, $k_{21} = 0.5$: Although this plant does certainly have non-zero off-diagonal elements, the Bristol index is $\lambda = 1.2$ and thus a strong coupling should not be expected. Moreover, the plant is still of minimum phase (zero at $s = -5$). The corresponding responses are shown in Figure 6.3 with solid lines. Here also, the closed loop could be easily speeded up. For example, multiplying the controller gains by factor five, a faster closed-loop response is attained with similar interaction amplitudes (but obviously requiring greater control action).

3. $k_{12} = 2$, $k_{21} = 1$: This is the original Rosenbrock system. Despite the simplicity of its individual transfer functions, the increase of the crossed gains makes the system to present non-minimum phase characteristics, turning its control harder. In Figure 6.4, it can be observed how the response obtained with the same controller (6.19) is considerably slower than in the previous cases. In addition, the interactions also have greater amplitude and settling time, in concordance with the larger Bristol index ($\lambda = 3$). In this case, if the gains K_p and K_i are increased the closed-loop response starts oscillating, and it becomes unstable from a factor greater than 2.5.

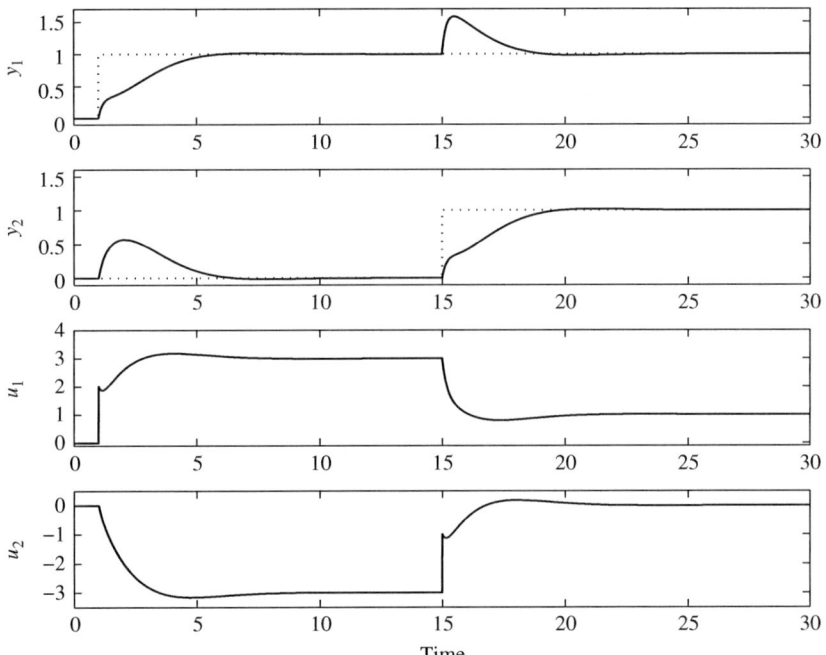

Figure 6.4   Closed-loop response of decentralised PI controller (6.19) and process (6.20) with $k_{12} = 2$ and $k_{21} = 1$ (Rosenbrock system)

4. $k_{12} = 1.7$, $k_{21} = 1.7$: The system remains stable, but the achieved responses look unacceptable, which is consistent with the large Bristol index ($\lambda = 27.27$) and the RHP zero too close to the origin ($s = 0.058$). The time

evolution of the outputs and control actions is presented in Figure 6.5, for which a scale change was performed and where the second loop step was moved to $t = 50$.

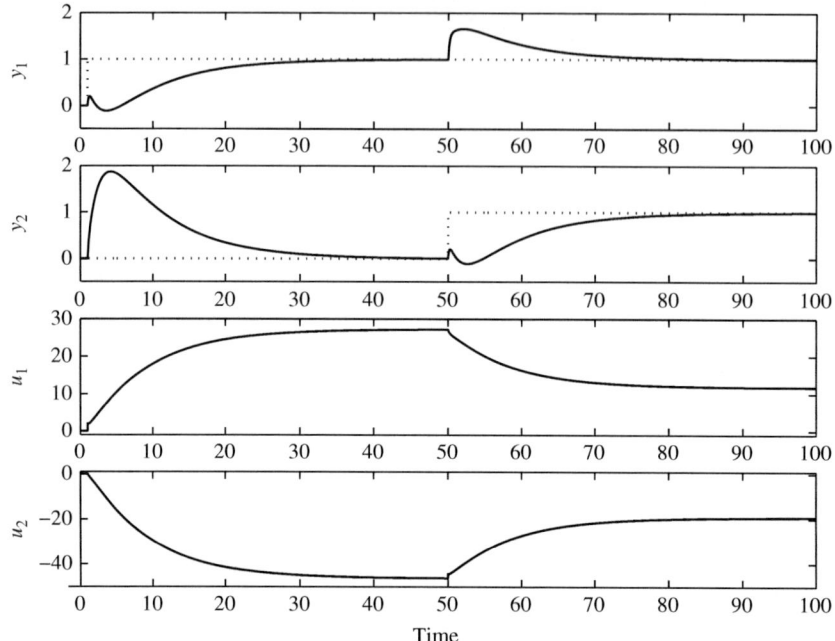

Figure 6.5   Closed-loop response of decentralised PI controller (6.19) and process (6.20) with $k_{12} = 1.7$ and $k_{21} = 1.7$

5. $k_{12} = 2$, $k_{21} = 2$: With these gains, the system becomes again of minimum phase ($z_0 = s = -0.33$). However, the closed loop with controller (6.19) results unstable. The Bristol interaction index is negative ($\lambda = -3$), which suggests permuting the controller connection. In fact, by interchanging the plant outputs the closed loop becomes stable, though excessively slow (see Figure 6.6).

As can be seen, once the closed-loop system is stabilised, the crossed interactions inherent to multivariable systems constitute one of the main performance limitations of decentralised control approaches. In Section 6.3, we present a useful technique based on SMRC and system state information to confine interactions to a desired range of amplitudes. Even then, in several control problems it will be necessary to consider centralised control implementation in order to improve the closed-loop behaviour. In effect, the example presented in this section shows that despite its practical advantages, decentralised or multiloop control does not always lead to good responses in the controlled variables. This fact is particularly true for NMP systems, which encouraged us to think about an alternative centralised strategy for the control of this type of systems, to be treated in next chapter.

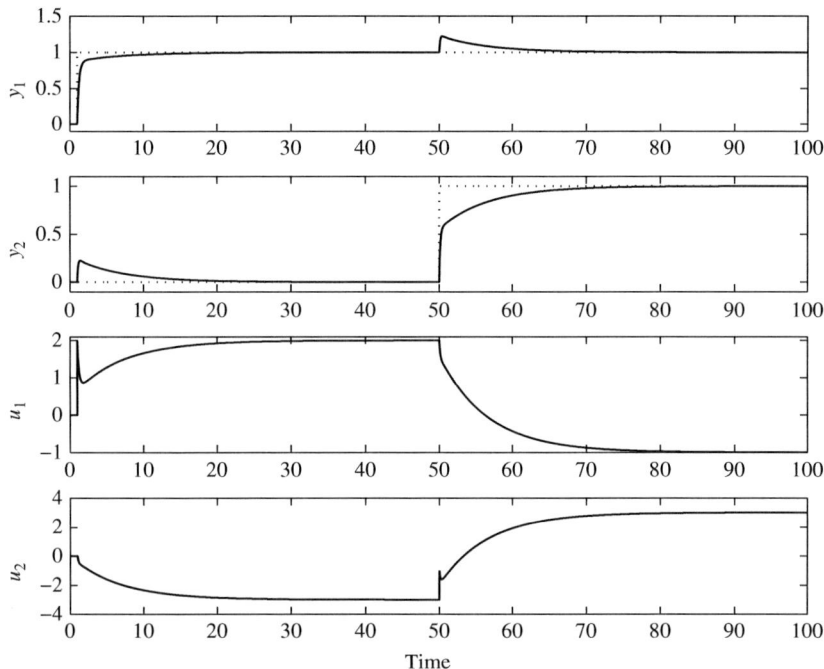

Figure 6.6 Closed-loop response of decentralised PI controller (6.19) and process (6.20) with interchanged outputs and $k_{12} = 2$ and $k_{21} = 2$

6.3 Limiting interactions in decentralised control via SMRC

In this section, we present a method to delimit the amplitude of crossed interaction in a multiloop control system [38]. The methodology is based on the SMRC technique described in Chapter 2. The objective here will be to impose predefined bounds to the interactions between the loops of the decentralised control system.

The method guarantees that the outputs that should ideally remain unchanged (whose set points did not change) keep inside a given range of values, which can be predefined as a design parameter. Hence, not only does SMRC improve the decoupling of the system, but it also enables a safe operation mode.

Although the algorithm does not present any limitation with respect to the number of inputs and outputs, the presentation is restricted to systems with two inputs and two outputs for clarity sake. The extension to greater dimensions is straightforward.

6.3.1 Control scheme

The dotted box in Figure 6.7 illustrates a decentralised control of a TITO process. An auxiliary loop for the conditioning of the reference signals has been added to this control scheme in order to delimit crossed interactions.

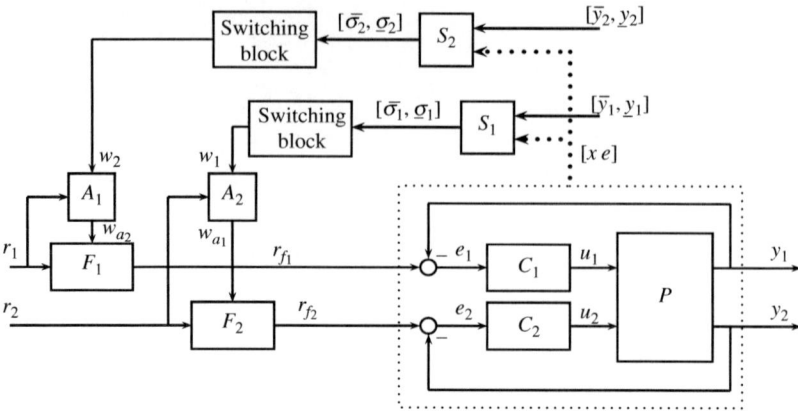

Figure 6.7   Decentralised control system with auxiliary conditioning loop for interaction limitation

C_1 and C_2 are realisable SISO controllers tuned according to the desired performance. It is assumed that the control pairing has been correctly chosen in a previous design from some interaction measures. For the sake of simplicity, it is also supposed that references r_1 and r_2 do not change simultaneously.[1] P represents the process under control which may be either stable or unstable. Outside the dotted box is the SMRC loop to limit interactions. It is composed of the following:

- S_1 (S_2): therein sliding mode functions $\underline{\sigma}_1$ and $\bar{\sigma}_1$ ($\underline{\sigma}_2$ and $\bar{\sigma}_2$) are computed taking into account the lower and the upper interaction limits $\underline{y}_1$ and $\bar{y}_1$ ($\underline{y}_2$ and $\bar{y}_2$), respectively.
- Switching blocks: they give rise to discontinuous signals w_1 and w_2 according to the sliding functions values.
- Logic block A_1 (A_2): it enables/disables the shaping of changed/unchanged reference r_1 (r_2), by making $w_{a2} = w_2$ or $w_{a2} = 0$ ($w_{a1} = w_1$ or $w_{a1} = 0$).
- F_1 and F_2: these are first-order reference filters, with cut-off frequencies much greater than the corresponding closed-loop bandwidths, in such a way that they do not affect by their own closed-loop dynamics.

Assume the model of the process P is known, having a minimal realisation in the form

$$P : \begin{cases} \dot{x}_p = A_p x_p + B_p u \\ y = C_p x_p + d_0 \end{cases} \qquad (6.24)$$

---

[1] Note that most practical problems are indeed excited by individual changes of the set points. However, this assumption is only made to simplify the method explanation since, as we will see in the next chapters, it is not strictly necessary.

wherein matrices A_p, B_p and C_p have dimensions $n \times n$, $n \times 2$ and $2 \times n$, respectively. The input is $u = [u_1 \ u_2]^T$ and the output is $y = [y_1 \ y_2]^T$. Output disturbances are represented by $d_0 = [d_1 \ d_2]^T$.

The decentralised controller pair $C_1 - C_2$ has the following state-space realisation:

$$C : \begin{cases} \dot{x}_c = A_c x_c + B_c e \\ u = C_c x_c + D_c e \end{cases} \tag{6.25}$$

where A_c, B_c and C_c are block-diagonal matrices of appropriate dimensions, D_c is a 2×2 diagonal matrix and $e = [e_1 \ e_2]^T$.

Therefore, since $e = r_f - y$, the closed-loop dynamics from the reference r_f to the process output y is given by

$$\dot{x} = Ax + B r_f + B_d d$$
$$y = Cx + d \tag{6.26}$$

where $x = [x_p \ x_c]^T$ and

$$A = \begin{bmatrix} A_p - B_p D_c C_p & B_p C_c \\ -B_c C_p & A_c \end{bmatrix}, \quad B = -B_d = \begin{bmatrix} B_p D_c \\ B_c \end{bmatrix}, \quad C = [C_p \ 0] \tag{6.27}$$

Besides, the dynamic behaviour of the filters $F_1 - F_2$ is described in a compact form as

$$F : \begin{cases} \dot{x}_f = A_f x_f + B_f r + B_w w_a \\ r_f = C_f x_f \end{cases} \tag{6.28}$$

with $A_f = -C_f = diag(\lambda_{f_1}, \lambda_{f_2})$, $B_f = I_2$ and B_w being the 2×2 permutation matrix. Also, $w_a = [w_{a1} \ w_{a2}]^T$, $r_f = [r_{f_1} \ r_{f_2}]^T$ and $r = [r_1 \ r_2]^T$.

6.3.2   Switching law

In order to delimit the closed-loop interactions by conditioning the reference r_{f_j} ($j = 1, 2$), the discontinuous control signal w_i ($i = 1, 2$; $i \neq j$) is governed by the following switching law (implemented in the switching block):

$$\begin{cases} w_i = w_i^- & \text{if} \quad \overline{\sigma}_i < 0 \\ w_i = w_i^+ & \text{if} \quad \underline{\sigma}_i > 0, \quad i = 1, 2 \\ w_i = 0 & \text{otherwise} \end{cases} \tag{6.29}$$

with

$$\overline{\sigma}_i(x, e_j, \overline{y}_i) = \overline{\psi}_i(\overline{y}_i) - \gamma_i^T x - \kappa_{ij} e_j$$
$$\underline{\sigma}_i(x, e_j, \underline{y}_i) = \underline{\psi}_i(\underline{y}_i) - \gamma_i^T x - \kappa_{ij} e_j \tag{6.30}$$

The non-zero scalar κ_{ij} and the row vector of n elements γ_i^T determine the output dynamics of the conditioning loop. $\tilde{\psi}_i$ are constant values that depend on interaction limits $\tilde{y}_i$ (recall $\tilde{\ast}$ stands for either $\bar{\ast}$ or $\underline{\ast}$).

Interaction limits $\tilde{y}_i$ can be defined by the control designer either as absolute constant values (usually for safe operation mode) or as relative to reference value r_i. In the last case, $\bar{y}_i = r_i + \delta_i^+$ and $\underline{y}_i = r_i - \delta_i^-$, where $[-\delta_i^-, \delta_i^+]$ is the range of variation allowed for y_i when r_j changes.

Note that the explicit presence of $e_j = r_{f_j} - y_j$ in (6.30) guarantees that the sliding functions have relative degree of one with respect to w_i, i.e. they satisfy the transversality condition. Effectively, $\dot{r}_{f_j}$ depends on w_i because F_j is a first-order filter.

Now, the following choice is made so that (6.30) can be written in terms of the canonical closed-loop states [38]:

- $\kappa_{ij} = k_{\rho_{ij}+1}^i c_i A^{\rho_{ij}-1} b_j$, c_i being the ith row of C, b_j the jth column of B and ρ_{ij} the relative degree of the transfer function between the filtered reference r_{f_j} and the process output y_i.

- $\gamma_i^T = \Gamma_i^T \cdot O_{(i,\rho_{ij}+1)} + \Phi_i^T$, where $\Gamma_i = [k_1^i \ \ldots \ k_{\rho_{ij}+1}^i]^T$ is a vector containing constant gains with its first element $k_1^i = 1$, and $\Phi_i = col(0, \ldots, 0, \kappa_{ij}c_j)$. Also,

$$O_{(i,\rho_{ij}+1)} = \begin{bmatrix} c_i \\ c_i A \\ \ldots \\ c_i A^{\rho_{ij}} \end{bmatrix} \tag{6.31}$$

comprises the first $\rho_{ij} + 1$ rows of the observability matrix of the subsystem with output y_i.

- $\tilde{\psi}_i(\tilde{y}_i) = \tilde{y}_i - k_{\rho_{ij}+1}^i \mathcal{C}$, where the constant $\mathcal{C} = c_i A^{(\rho_{ij}-1)} b_i r_{f_i}$ equals zero if $\rho_{ij} \neq \rho_{ii}$.

Therefore, sliding functions $\bar{\sigma}_i(x, e_j, \bar{y}_i)$ and $\underline{\sigma}_i(x, e_j, \underline{y}_i)$ can be rewritten in a more intuitive form (in terms of the outputs and their derivatives):

$$\tilde{\sigma}_i(x, e_j, \tilde{y}_i) = \tilde{y}_i - y_i - \sum_{\alpha=1}^{\rho_{ij}} k_{\alpha+1}^i y_i^{(\alpha)} \tag{6.32}$$

As seen in Chapter 2, this form is also useful to calculate the output dynamics during sliding regime.

Again, the discontinuous signal w_i switches between w_i^- and 0 from one side to the other of the surface $\bar{\sigma}_i = 0$, and between w_i^+ and 0 on both sides of the surface $\underline{\sigma}_i = 0$. Therefore, the switching law (6.43) states that in the region between the surfaces $\bar{\sigma}_i = 0$ and $\underline{\sigma}_i = 0$ trajectories evolve with the own direction of the system, but just beyond this linear region the trajectories are forced to change their direction. In this way, sliding mode is only established when there exists risk of exceeding interaction limits.

Since the objective here is to compensate for interactions caused by reference changes, as a rule of thumb it is reasonable to take $w_i^{\pm}$ of the order of the set-point changes (taking into account limits on output disturbances). Nevertheless, the minimum values for w_i^+ and w_i^- can be explicitly computed from the necessary and sufficient condition for sliding mode (SM) establishment.

$$w_i^- \leq w_{eq_i} \leq w_i^+ \tag{6.33}$$

where w_{eq_i} represents the continuous equivalent control. It can be obtained differentiating once (6.30) or (6.32) with respect to time and equaling zero. This yields

$$w_{eq_i} = [\kappa_i \lambda_{f_j}]^{-1} \{\gamma_i^T A x + (\gamma_i^T b_j + \kappa_i \lambda_{f_j}) r_{f_j} - \gamma_i^T b_j d_j - \kappa_i (\lambda_{f_j} r_j + \dot{y}_j) + \gamma_i^T b_i r_{f_i}\} \tag{6.34}$$

From (6.33) and (6.34), values of $w_i^\pm$ can be derived to assure SM on certain domains of the sliding surfaces (which are determined by reasonable bounds on x, r_j, d_j and $\dot{d}_j$), thus preventing the process output y_i from exceeding its coupling limits. Recall that the selection of $w_i^\pm$ can be made in a conservative manner because the SM is restricted to the low-power side of the system.

6.3.3 Output dynamics during conditioning

Since all real plants are represented by strictly proper models, and all realisable controllers are proper (in particular P, PI and realisable PID controllers), the transfer function between y_i and r_{f_j} will always be strictly proper ($\rho_{ij} \geq 1$). Hence, the dynamic analysis can be performed following the development of Section 2.5.

In fact, considering in (6.26) and (6.27) a single output y_i and regarding that only input r_{f_j} and disturbance d_j change, we obtain the following normal form, which includes the filter dynamics:

$$\begin{cases} \dot{z}_1 &= z_2 \\ \dot{z}_2 &= z_3 \\ \cdots &= \cdots \\ \dot{z}_{\rho_{ij}-1} &= z_{\rho_{ij}} \\ \dot{z}_{\rho_{ij}} &= a_\xi \xi_s + a_\eta \eta_s + c_i A^{(\rho_{ij}-1)} b_j \, (r_{f_j} - d_j) + \mathcal{C} \\ \dot{\eta}_s &= P\xi_s + Q\eta_s \\ \dot{r}_{f_j} &= \lambda_{f_j}(r_{f_j} - r_j - w_{ai}) \end{cases} \tag{6.35}$$
$$y_i = z_1$$

where $\xi_s = [z_1 \, z_2 \, \ldots \, z_{\rho_{ij}}]^T$ comprises the process output y_i and its first $(\rho_{ij} - 1)$ derivatives, $\eta_s = [z_{\rho_{ij}+1} \, \ldots \, z_{n_{ij}}]^T$ are $(n_{ij} - \rho_{ij})$ linearly independent states (n_{ij} being the order of the transfer function between y_i and r_{f_j}) and $c_i A^{(\rho_{ij}-1)} b_j$ is a non-zero constant. Recall that the zeros of the transfer function between r_{f_j} and y_i are the eigenvalues of Q. Then, for the hidden dynamics to be stable, the off-diagonal elements of the closed-loop transfer matrix $T(s)$ should not have RHP zeros. However, it is not necessary for the process under control to be of minimum phase, since transmission zeros may be in the RHP (as in the example to be presented).

Rewriting now (6.32) as a function of the states $x_{cl} = [\xi_s \ \eta_s \ r_{f_j}]^T$ of the representation (6.35) and equaling zero both sliding functions, we arrive at the following expression for the sliding surfaces:

$$\tilde{\sigma}_i = \tilde{y}_i - y_i - \sum_{\alpha=2}^{\rho_{ij}} k_\alpha^i z_\alpha - k_{\rho_{ij}+1}^i (a_\xi \xi_s + a_\eta \eta_s + c_i A^{(\rho_{ij}-1)} b_j (r_{f_j} - d_j) + C) = 0$$

$$(6.36)$$

Thus, (6.35) becomes redundant during SM. Indeed, (6.36) establishes a linear relationship among $(r_{f_j} - d_j)$ and the states ξ_s and η_s, that if replaced in (6.35) yields

$$\begin{cases} \dot{z}_1 &= z_2 \\ \dot{z}_2 &= z_3 \\ \cdots &= \cdots \\ \dot{z}_{\rho_{ij}-1} &= z_{\rho_{ij}} \\ \dot{z}_{\rho_{ij}} &= (\tilde{y}_i - y_i - \sum_{\alpha=2}^{\rho_{ij}} k_\alpha^i z_\alpha)/k_{\rho_{ij}+1}^i \\ \dot{\eta} &= P\xi + Q\eta \end{cases} \qquad (6.37)$$

Then, the dynamics with which y_i evolves towards its limit $\tilde{y}_i$ during SMRC does only depend on the values k_α^i chosen for the switching functions design. Consequently, the control system designer can easily delimit the interaction amplitude to $|\tilde{y}_i - r_i|$, with $\tilde{y}_i$ being a design parameter.

6.3.4 Behaviour in presence of output disturbances

We consider now output disturbances in the jth loop, in order to analyse how the SMRC method behaves with the coupling effects of d_j on output y_i. With this objective, the state equation of the conditioning loop can be calculated from (6.28) and (6.26)–(6.27), yielding

$$\begin{bmatrix} \dot{x} \\ \dot{e}_j \end{bmatrix} = \begin{bmatrix} A + b_j c_j & b_j \\ 0^T & \lambda_{f_j} \end{bmatrix} \begin{bmatrix} x \\ e_j \end{bmatrix} + \begin{bmatrix} b_i r_{f_i} \\ 0 \end{bmatrix} + \begin{bmatrix} 0 \\ \lambda_{f_j}(y_j - r_j) - \dot{y}_j \end{bmatrix} + \begin{bmatrix} 0 \\ \lambda_{f_j} w_{a_i} \end{bmatrix}$$

$$(6.38)$$

Clearly, the second term of the right-hand side of (6.38) is constant, whereas the last two terms can be interpreted as the disturbance and control vector fields of the conditioning loop, respectively. So, the disturbance satisfies the *matching condition*, i.e. $[0 \ \lambda_{f_j}(y_j - r_j) - \dot{y}_j]^T \in span([0 \ \lambda_{f_j} w_{a_i}]^T)$. Then, as can be verified in (6.37), the conditioning dynamics exhibits strong invariance with respect to the output disturbance d_j and its derivative $\dot{d}_j$ (included in y_j and $\dot{y}_j$) provided the condition (6.33) holds. This latter condition requires d_j and $\dot{d}_j$ to be bounded, which is always true in practice.

Therefore, for bounded and smooth output disturbances d_j the dynamics of the output y_i is completely governed by the feedback gains k_a^i whenever the conditioning loop is active.

Remark 6.3: *As we have seen in the previous chapters, loop interactions can be eliminated by means of a centralised decoupling controller. However, for the great number of processes in industry which are already under decentralised control structures, this would imply to completely change the control system. The SMRC method allows one to delimit loop interactions without having to change neither the overall control system nor the single-loop existing controllers. Naturally, due to the limitations inherited from decentralised control, a closed-loop performance like the one that could be achieved by centralised control should not be expected.*

6.4 Two-degrees of freedom PID controller with adaptive set-point weighting

Among the large number of modifications to the standard structure of PID control proposed with the aim of improving closed-loop performance, one of the most widely accepted is the so-called PID with set-point weighting or two-degrees of freedom PID (2DOF-PID), which is shown in Figure 6.8. The 2DOF-PID structure allows adjusting separately the response of the system to disturbances and to set-point changes.

This alternative PID structure has also been exploited to deal with the interaction of MIMO systems. In Reference 3, it is shown that the interaction measures are minimised by setting the set-point weighting factor equal to zero. That is, it is claimed that the best configuration in terms of coupling reduction is the so-called I–P structure. However, cancelling the set-point weighting leads, at the same time, to slower closed-loop responses.

To tackle this compromise, the set-point weighting of 2DOF-PID can be adapted using the SMRC algorithm of Section 6.3, rather than fixing it to a constant value, as is generally done.

A 2DOF-PI controller for TITO systems can be represented as

$$u(t) = K_p e_p(t) + K_i \int_0^t e(t)\, dt \tag{6.39}$$

where

$$e_p(t) = Br(t) - y(t) \tag{6.40}$$

$$e(t) = r(t) - y(t) \tag{6.41}$$

where $r(t) \in \mathbb{R}^2$ and $y(t) \in \mathbb{R}^2$ are the set-point or reference vector and the controlled variable vector, respectively. K_p and K_i are the diagonal constant matrices containing the controller gains, while $B \in \mathbb{R}^{2 \times 2}$ is also a diagonal and is composed of the set-point weighting factors for each loop, as shown in Figure 6.8.

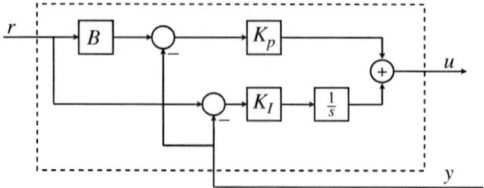

Figure 6.8   PI controller with set-point weighting

Let us suppose without loss of generality that our objective is to bound the transient amplitude on the controlled variable y_2 when a step change is applied to set point r_1 (and r_2 remains constant). With this purpose, two switching functions $\bar{\sigma}(x)$ and $\underline{\sigma}(x)$ can be constructed from the system states as follows:

$$\tilde{\sigma}(x, u, \tilde{y}_2) = \tilde{y}_2 - \kappa_2^{\mathrm{T}}[x_p \; u]^{\mathrm{T}} \tag{6.42}$$

where $x_p \in \mathbb{R}^n$ is the system state and $\kappa_2 \in \mathbb{R}^{n+2}$ is the constant vector used to reconstruct ρ_{21} consecutive derivatives of y_2 from the system states.

According to the sign of these functions, a scalar parameter b_{1d} is then set as follows:

$$\begin{cases} b_{1d} = 0 & \text{if} \quad \bar{\sigma}(x, u, \bar{y}_2) < 0 \\ b_{1d} = 1 & \text{if} \quad \underline{\sigma}(x, u, \underline{y}_2) > 0 \\ b_{1d} = b_{1_0} & \text{otherwise} \end{cases} \tag{6.43}$$

where b_{1_0} is the value chosen for set-point weighting when interaction amplitude limits on y_2 are not surpassed (i.e. b_{1_0} is the default weighting). The adaptive set-point weighting b_1 is finally obtained by passing b_{1d} through a first-order low-pass filter.

Note that both the design and analysis of this augmented 2DOF-PID control are the same as those of the previous section, with the difference that only the proportional path of the reference is now conditioned. The above procedure allows improving closed-loop step responses achieved by I–P configurations for a given interaction tolerance (simulation results in presence of measurement noise and model uncertainty can be found in Reference 42). Furthermore, interaction tolerance can be directly included as a design parameter, which has the advantage of being extremely intuitive for the control system operator.

6.5   Case study: Quadruple tank

We consider now a multivariable process with an adjustable transmission zero which is known as *quadruple tank*. It was originally presented in Reference 66, and from then it has been extensively used as a benchmark system to assess the performance of different MIMO control strategies. Figure 6.9 presents a schematic diagram of the quadruple tank.

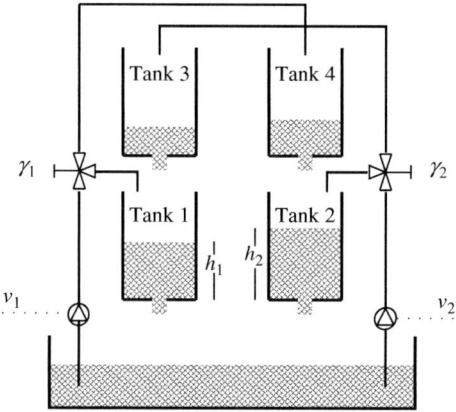

Figure 6.9 Schematic diagram of the quadruple-tank system

In the first place, we will examine the main properties of this laboratory-scaled plant. We are interested in analysing how some of the studied MIMO features, such as the transmission zeros or RGA, depend on the system parameters. Subsequently, the SMRC algorithm is applied for a setting in which the system is very hard to control and decentralised control leads to excessively large interactions.

6.5.1 Plant model analysis

Let us consider the following non-linear model for the quadruple tank, derived from physical data, mass balances and Bernoulli's law:

$$\frac{dh_1}{dt} = -\frac{a_1}{A_1}\sqrt{2gh_1} + \frac{a_3}{A_1}\sqrt{2gh_3} + \frac{\gamma_1 k_1}{A_1}v_1$$

$$\frac{dh_2}{dt} = -\frac{a_2}{A_2}\sqrt{2gh_2} + \frac{a_4}{A_2}\sqrt{2gh_4} + \frac{\gamma_2 k_2}{A_2}v_2$$

$$\frac{dh_3}{dt} = -\frac{a_3}{A_3}\sqrt{2gh_3} + \frac{(1-\gamma_2)k_2}{A_3}v_2 \tag{6.44}$$

$$\frac{dh_4}{dt} = -\frac{a_4}{A_4}\sqrt{2gh_4} + \frac{(1-\gamma_1)k_1}{A_4}v_1$$

where h_i represents the water level in each tank. A_i and a_i are the cross-sections of the tanks and the outlet holes, respectively. The constants γ_1, $\gamma_2 \in (0,1)$ are determined from how two flow divider valves are set. The process outputs (y_1, y_2) are the signals in volts generated by the sensors in the lower tanks ($y_1 = k_c h_1$ and $y_2 = k_c h_2$), while the inputs to the system are the voltages v_i applied to the two pumps (the corresponding flow is $q_i = k_i v_i$).

The system parameter values are given in Table 6.1 for the two operating points considered in Reference 66: $P-$, at which the system presents minimum phase characteristics, and $P+$, where it behaves as a non-minimum phase process.

Table 6.1   Quadruple-tank parameters for MP (P−) and NMP (P+) configurations

Fixed parameters	Notation	Value
Cross-section (tanks)	A_1, A_3	28 cm^2
	A_2, A_4	32 cm^2
Cross-section (outlet holes)	a_1, a_3	0.071 cm^2
	a_2, a_4	0.057 cm^2
Sensors sensitivity	k_c	0.5 V/cm
Acceleration of gravity	g	981 cm/s
Parameters at P−	**Notation**	**Value at P−**
Tanks levels	$(h_1^0; h_2^0)$	(12.4; 2.7) cm
	$(h_3^0; h_4^0)$	(1.8; 1.4) cm
Pumps input voltage	$(v_1^0; v_2^0)$	(3.00; 3.00) V
Flow/voltage constant	$(k_1; k_2)$	(3.33; 3.35) cm^3/V s
Valves position	$(\gamma_1; \gamma_2)$	(0.7; 0.6)
Parameters at P+	**Notation**	**Value at P+**
Tanks levels	$(h_1^0; h_2^0)$	(12.6; 12.4) cm
	$(h_3^0; h_4^0)$	(4.8; 4.9) cm
Pumps input voltage	$(v_1^0; v_2^0)$	(3.15; 3.15) V
Flow/voltage constant	$(k_1; k_2)$	(3.14; 3.29) cm^3/V s
Valves position	$(\gamma_1; \gamma_2)$	(0.43; 0.34)

Defining variables $x_i = h_i - h_i^0$ and $u_i = v_i - v_i^0$, the linearised process model around a generic operating point has the following expression in the state space:

$$\frac{dx}{dt} = \begin{bmatrix} -\dfrac{1}{T_1} & 0 & \dfrac{A_3}{A_1 T_3} & 0 \\ 0 & -\dfrac{1}{T_2} & 0 & \dfrac{A_4}{A_2 T_4} \\ 0 & 0 & -\dfrac{1}{T_3} & 0 \\ 0 & 0 & 0 & -\dfrac{1}{T_4} \end{bmatrix} x + \begin{bmatrix} \dfrac{\gamma_1 k_1}{A_1} & 0 \\ 0 & \dfrac{\gamma_2 k_2}{A_2} \\ 0 & \dfrac{(1-\gamma_2)k_2}{A_3} \\ \dfrac{(1-\gamma_1)k_1}{A_4} & 0 \end{bmatrix} u \tag{6.45}$$

$$y^\delta = \begin{bmatrix} k_c & 0 & 0 & 0 \\ 0 & k_c & 0 & 0 \end{bmatrix} x$$

where

$$T_i = \frac{A_i}{a_i} \sqrt{\frac{2h_i^0}{g}}, \quad i = 1, \dots, 4 \tag{6.46}$$

are the time constants that for each operating point take the values as given below:

Time constant	$P-$	$P+$
(T_1, T_2)	$(62, 90)$ s	$(63, 91)$ s
(T_3, T_4)	$(23, 30)$ s	$(39, 56)$ s

In the frequency domain, the corresponding transfer matrix is

$$P(s) = \begin{bmatrix} \dfrac{\gamma_1 T_1 k_1 k_c}{A_1(1 + T_1 s)} & \dfrac{(1 - \gamma_2) T_1 k_1 k_c}{A_1(1 + T_1 s)(1 + T_3 s)} \\[4mm] \dfrac{(1 - \gamma_1) T_2 k_2 k_c}{A_2(1 + T_2 s)(1 + T_4 s)} & \dfrac{\gamma_2 T_2 k_2 k_c}{A_2(1 + T_2 s)} \end{bmatrix} \tag{6.47}$$

The zeros of this transfer matrix are the roots of the numerator polynomial of

$$det(P(s)) = \frac{\gamma_1 \gamma_2 T_1 T_2 k_1 k_2 k_c^2}{A_1 A_2 \prod_{i=1}^{4}(1 + s T_i)} \left[(1 + s T_3)(1 + s T_4) - \frac{(1 - \gamma_1)(1 - \gamma_2)}{\gamma_1 \gamma_2} \right] \tag{6.48}$$

Calling

$$k = \frac{(1 - \gamma_1)(1 - \gamma_2)}{\gamma_1 \gamma_2} \in (0, \infty) \tag{6.49}$$

the system zeros will be those values of s that verify

$$(1 + s T_3)(1 + s T_4) - k = 0 \tag{6.50}$$

That is, the system has two multivariable zeros. From a complementary root-locus analysis, it is easily derived that one of these zeros will always be in the LHP, whereas the other may be either in the LHP or in the RHP depending on the value of k (and thus, on γ_1 and γ_2).

For small values of k, the zeros will be close to $s = -1/T_3$ and $s = -1/T_4$. As k increases and tends to infinite, one zero tends to $+\infty$ and the other one to $-\infty$. Note from (6.50) that when $k = 1$, one of the zeros will be at the origin. Moreover, from (6.49) $k = 1$ implies that $\gamma_1 + \gamma_2 = 1$. Then, according to this reasoning

$$0 < \gamma_1 + \gamma_2 < 1 \rightarrow \text{NMP system}$$

$$1 < \gamma_1 + \gamma_2 < 2 \rightarrow \text{MP system}$$

These two regions are shown in Figure 6.10, divided by the solid straight line that joins the points $(\gamma_1, \gamma_2) = (0, 1)$ and $(\gamma_1, \gamma_2) = (1, 0)$. Physically, it is easy to show that – for the same flow in each pump – the system is NMP when the flow towards the upper tanks is greater than the flow towards the lower tanks ($\gamma_1 + \gamma_2 < 1$). Conversely, if the flow towards tanks 1 and 2 is greater than the flow towards tanks 3 and 4 ($\gamma_1 + \gamma_2 > 1$), the system behaves as an MP process [66].

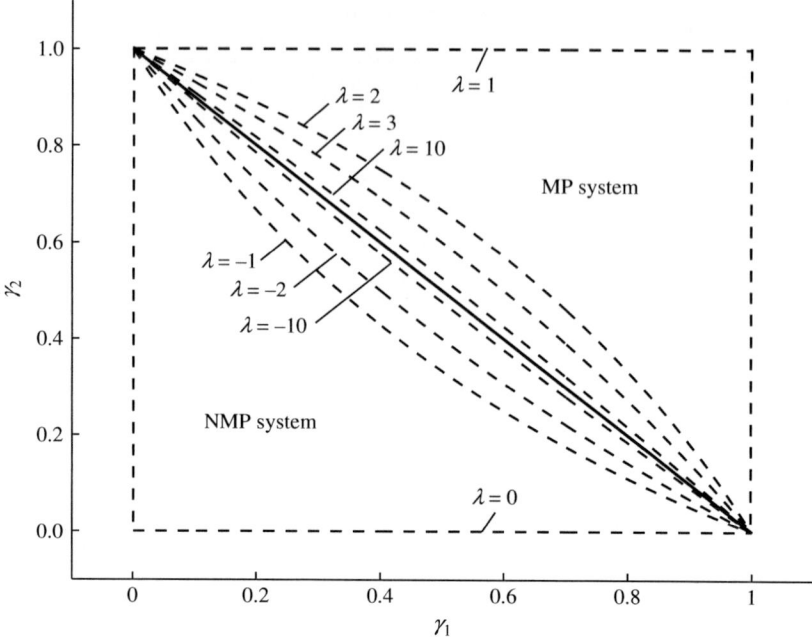

Figure 6.10   Quadruple-tank MP and NMP regions depending on the valves positions γ_1 and γ_2

Besides, the RGA of (6.47) is given by the Bristol interaction index

$$\lambda = \frac{p_{11}(0)p_{22}(0)}{p_{11}(0)p_{22}(0) - p_{12}(0)p_{21}(0)} = \frac{-\gamma_1\gamma_2}{1 - \gamma_1\gamma_2} \tag{6.51}$$

which only depends on the position of the flow divider valves.

The dashed curves given in Figure 6.10 show the geometric loci on the plane (γ_1, γ_2) corresponding to different values of the Bristol index. As $\gamma_1 + \gamma_2$ tends to 1 (solid line), $|\lambda|$ increases. Note also from (6.51) that there are no $\gamma_1, \gamma_2 \in [0, 1]$ such that $\lambda \in [0, 1]$.

The unfavourable situation $\lambda < 0$ occurs for $0 < \gamma_1 + \gamma_2 < 1$, i.e. for those configurations in which the system exhibits NMP characteristics, which is evident from the process physical interpretation. For such cases, we have seen that the input/output pairing should be modified in order to apply a decentralised control structure. If process outputs are interchanged, the Bristol interaction index yields

$$\lambda' = \frac{(1 - \gamma_1)(1 - \gamma_2)}{1 - \gamma_1 - \gamma_2} \tag{6.52}$$

Note that now $\lambda > 0$ for $0 < \gamma_1 + \gamma_2 < 1$.

6.5.2   Interactions limits in non-minimum phase setting

Replacing the parameters values at the operating point $P+$ in (6.47) yields

$$
P_+(s) = \begin{bmatrix} \dfrac{1.5}{1+63s} & \dfrac{2.5}{(1+39s)(1+63s)} \\[2ex] \dfrac{2.5}{(1+56s)(1+91s)} & \dfrac{1.6}{1+91s} \end{bmatrix} \tag{6.53}
$$

It is straightforward to verify that the transfer matrix $P_+(s)$ has a transmission zero at $s = 0.013$, while its RGA is given by

$$
\Lambda = \begin{bmatrix} -0.6356 & 1.6356 \\ 1.6356 & -0.6356 \end{bmatrix} \tag{6.54}
$$

Clearly, the RGA suggests that another input/output pairing should be chosen. This is verified by the great interactions that can be appreciated in the simulations run in Reference 66 with decentralised PI controllers and this original control structure.

Since it does not entail any additional effort, we first permute here the system outputs in order to reduce the interactions amplitude. The RGA is then given by the matrix which results from interchanging the first and second row of (6.54) (the Bristol interaction index now becomes greater than zero).

As explained in Reference 66, although the controllers' manual tuning is relatively simple for the MP configuration $(P-)$, finding suitable controllers for the NMP setting $(P+)$ is a much harder task. In effect, the PI controllers used in the cited reference for the original plant $P_+(s)$ (without output permutation) are not able to stabilise the permuted plant. Therefore, those controllers are retuned for the permuted plant, finding that the following decentralised PI controllers

$$
C_1(s) = 0.15\left(1 + \frac{1}{50s}\right), \quad C_2(s) = 0.25\left(1 + \frac{1}{30s}\right) \tag{6.55}
$$

give acceptable closed-loop responses. Compared with the responses obtained in Reference 66, the permuted multiloop architecture employed here achieves shorter settling times and smaller interaction amplitude. However, it is important to recall that the PI controller tuning is not the objective of this chapter. Indeed, this tuning may surely be improved by means of some advanced tuning techniques. What we aim here is to delimit the crossed interactions in decentralised control systems, whichever is the methodology used to design the controllers of each loop. It is illustrated next.

To evaluate the SMRC methodology, we add to the system the filters $f_1(s)$ and $f_2(s)$ like the ones described by (6.28). For this application, $f_1(s)$ and $f_2(s)$ were chosen identical, with eigenvalues $\lambda_{f_1} = \lambda_{f_2} = -0.5$.

The closed-loop system with controllers (6.55) was then excited with a unitary step in r_2 ($t = 50$ s), and subsequently with another unitary step in r_1 ($t = 2500$ s). The nominal closed-loop system response to such an input is shown in Figure 6.11, together with the reference signals r_1 and r_2 (thin-dotted line). As can be noticed,

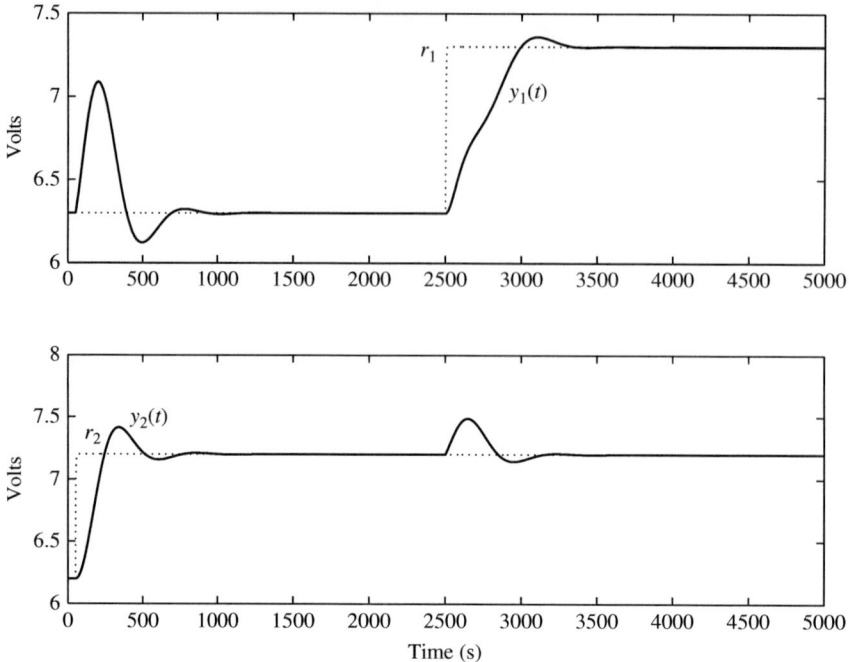

Figure 6.11   Output signals y under decentralised control and step references in
r₁ (t = 2500 s) and r₂ (t = 50 s)

the closed-loop system presents large interactions even for the permuted control structure, particularly from input r_2 to output y_1.

The SMRC technique is now applied with the aim of reducing these interactions. We consider interaction limits $\delta_1^+ = \delta_2^+ = \delta_1^- = \delta_2^- = 0.2$. The four sliding surfaces can then be written as follows:

$$\begin{cases} \overline{s_1} = r_1 + \delta_1^+ - c_1 x - 10\,c_1(A + b_2 c_2)x - 10\,c_1 b_2 e_2 = 0 \\ \underline{s_1} = r_1 - \delta_1^- - c_1 x - 10\,c_1(A + b_2 c_2)x - 10\,c_1 b_2 e_2 = 0 \end{cases} \tag{6.56}$$

$$\begin{cases} \overline{s_2} = r_2 + \delta_2^+ - c_2 x - 10\,c_2(A + b_1 c_1)x - 10\,c_2 b_1 e_1 = 0 \\ \underline{s_2} = r_2 - \delta_2^- - c_2 x - 10\,c_2(A + b_1 c_1)x - 10\,c_2 b_1 e_1 = 0 \end{cases} \tag{6.57}$$

where the gains $k_2^1 = 10$ and $k_2^2 = 10$ were chosen for a time constant of approximately 10 s during reference conditioning. Also, A, B and C are the matrices of the state-space representation of the closed loop, with c_i (b_i) being the ith row (column) of the matrix C (B). The discontinuous signals w_1 and w_2 are generated from these sliding functions in the switching blocks. They switch according to the law (6.43), with $w_1^- = w_2^- = -1$ and $w_1^+ = w_2^+ = 1$.

The results obtained with the SMRC method are shown by solid line in Figure 6.12 (the original closed-loop responses are repeated in dotted lines for

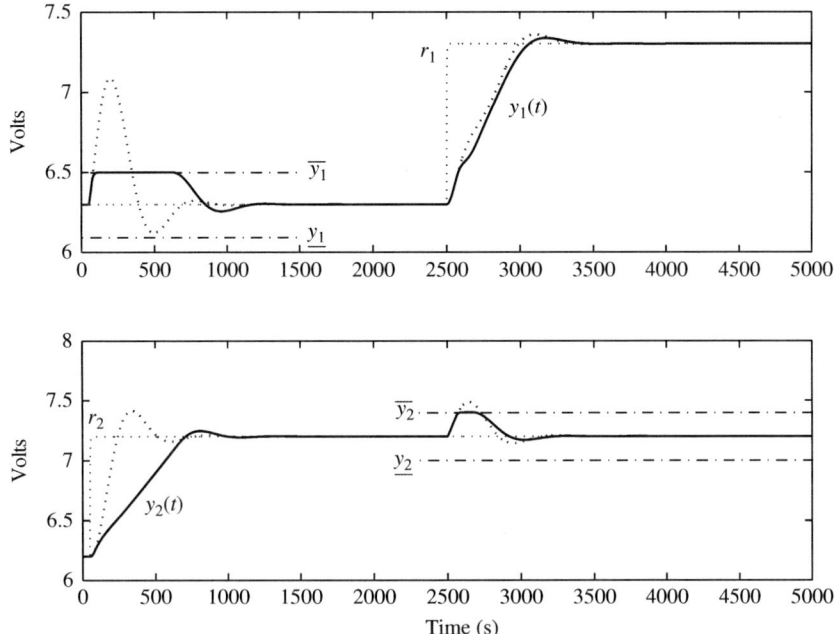

Figure 6.12   Output signals y under decentralised control with SMRC, for the same step references as given in Figure 6.11

comparative purposes). As can be seen, the interactions are bounded to the pre-defined limits, reducing indeed the coupling effects. Figure 6.13 shows the time evolution of the discontinuous conditioning signals, the filtered references and the control actions corresponding to Figure 6.12. After the step in reference r_2, a sliding regime is established along the surface $\overline{s_1} = 0$ just before y_1 reaches the upper interaction limit. Since then, the process output y_1 tends towards its limit $\overline{y_1}$ with the dynamics imposed by the constant gain k_2^1 of $\overline{s_1} = 0$. This SM shapes reference r_2 so that greater interaction is prevented, until y_1 returns to the linear region. Afterwards, a sliding mode establishes again after the step in r_1, now for a shorter period (see $w_{a2}(t)$ commutation) and on the surface $\overline{s_2} = 0$. Again, the limitation of interaction is evidenced.

Observe that, although it is not shown in the figures, when smaller set-point changes are performed (or smaller interactions occur) the system with SM compensation behaves exactly the same as the original system (without SM correction loop).

We are also interested in evaluating how the method manages to limit the crossed effects of output disturbances. This is illustrated by the simulation results presented in Figures 6.14 and 6.15.

Left boxes of Figure 6.14 depict the output transients caused by an output disturbance on y_2. Therein, the limitation of the coupling effect on output y_1 is verified. Again, solid line stands for SM case and dotted line stands for nominal

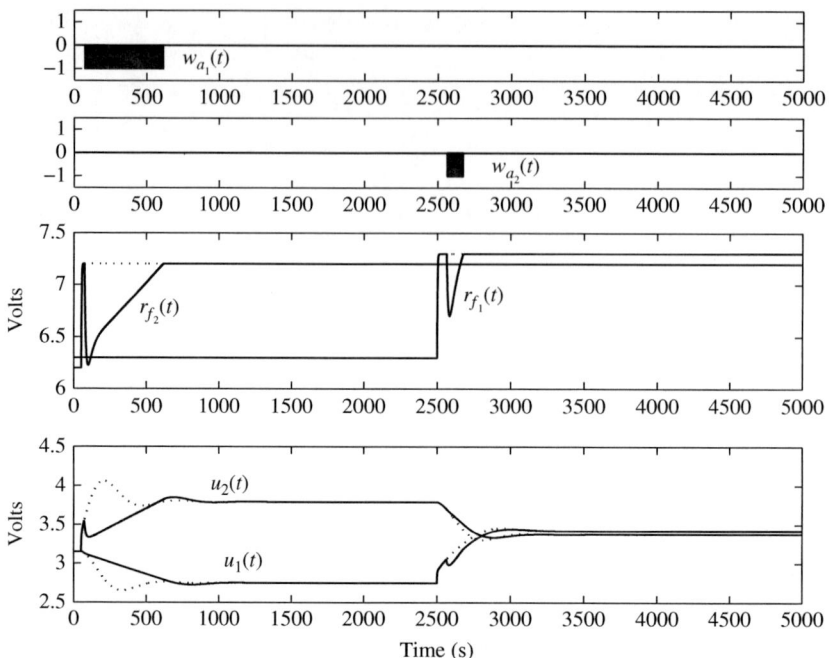

Figure 6.13   Discontinuous signals w_a after logic blocks, filtered references r_f and controller outputs u with (solid line) and without (dotted line) SM compensation, corresponding to Figure 6.12

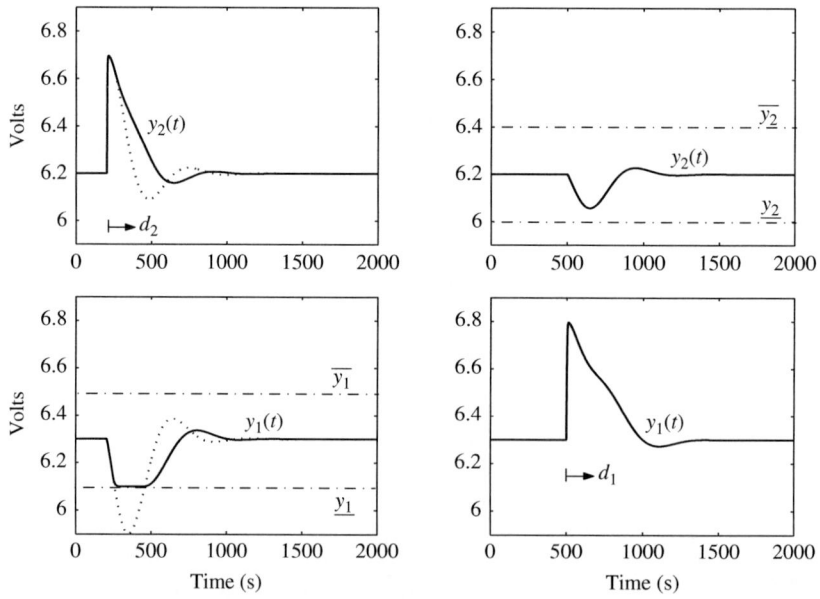

Figure 6.14   Output signals y with (solid line) and without (dotted line) SMRC when facing crossed output disturbances

closed loop. In turn, the right-hand part of the figure reveals for the case of an output disturbance on y_1 that the compensated system operates exactly the same as the original system when interaction limits are not reached (solid and dotted lines are overlapped).

In Figure 6.15, there is a plot of the output signals when both set-point changes and output disturbances occur in y_2. Even for the worst case at $t = 2700$, when positive disturbance in y_2 would have produced a greater interaction (see Figure 6.14), y_1 is kept between the chosen boundaries. The bottom half of the figure shows the output trajectories evolution on the plane (y_1, y_2) for the nominal closed loop (dotted line) and the SM compensated system (solid line). Differing from the latter, the original closed-loop trajectory largely exceeds the interaction limits.

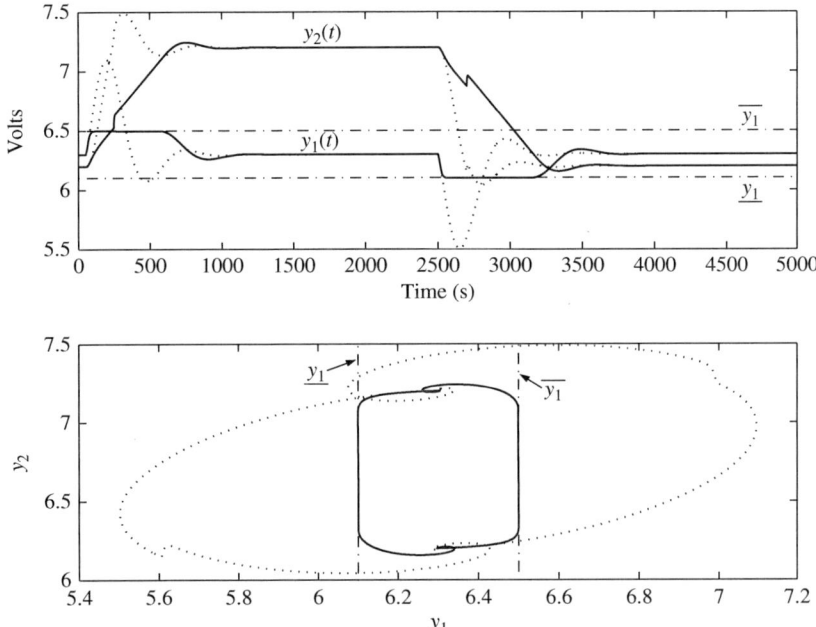

Figure 6.15   Process outputs y for step reference r_2 and disturbance d_2 (top plot), and the corresponding output trajectories (bottom plot)

As can be expected, the improvement achieved by the SMRC technique with respect to crossed coupling has naturally a potential cost in other performance indexes, which depends on the limit imposed to the interaction. Figure 6.16 analyses this cost for the quadruple-tank process.

The top plot in Figure 6.16 compares the settling time (3%) obtained with the original closed loop versus the one achieved by the SMRC method when different interaction limits are set. The tracking case given in Figure 6.12 was considered for this comparison. In particular, the worst case of r_2 tracking was chosen. Obviously, the original closed loop has always the same settling time T_{s0} (horizontal dotted line) as it

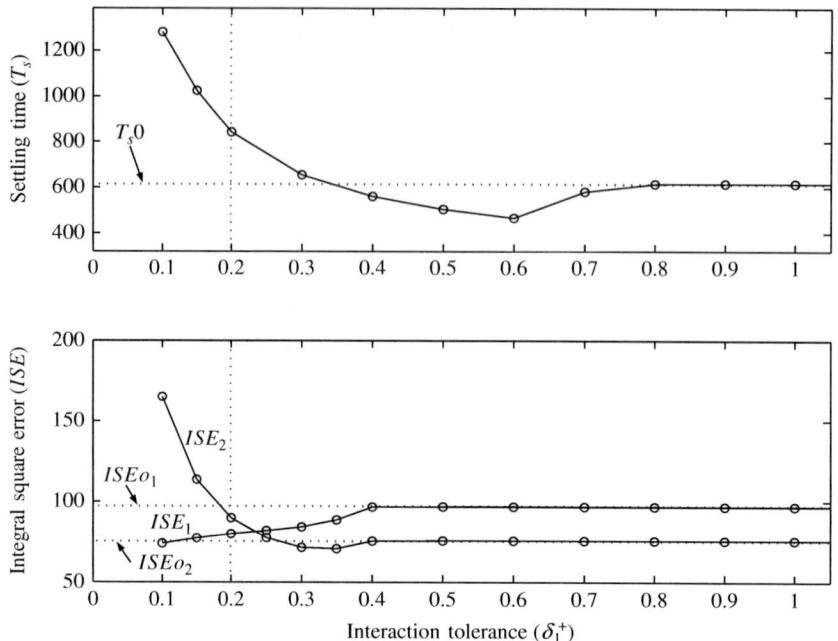

*Figure 6.16   (Upper plot) Settling time for variable y_2 given in Figure 6.12.
(Lower plot) Integral Square Error corresponding to the left part of
Figure 6.14*

does not include the compensation loop. When interaction is not hardly bounded, there is a range for which the SM compensated system has shorter settling time than the original system (up to approximately 70% interaction reduction). This should not be expected in general, since it is not the aim of the method. As the upper interaction limit δ_1^+ decreases beyond 0.3, the settling time increases. The vertical dotted line indicates the case considered in previous figures, which can be verified in Figure 6.12.

Integral Square Error (*ISE*), usually used as a regulation performance index, is considered in the lower box shown in Figure 6.16. Therein, *ISE* is plotted for the perturbed case shown in Figure 6.14 (left boxes), but varying the tolerance to the disturbance crossed effect. Again, original closed-loop *ISEo*$_1$ and *ISEo*$_2$ are depicted in horizontal dotted lines, and the case considered in the left boxes shown in Figure 6.14 is indicated with a vertical dotted line. Although there is a region for which both *ISE*$_1$ and *ISE*$_2$ are lesser than the nominal ones, for δ_1^+ inferior to 0.25 it is evidenced how *ISE*$_1$ improves while *ISE*$_2$ deteriorates. It seems logical because *ISE*$_1$ corresponds to the limited output y_1, but *ISE*$_2$ is calculated from the perturbed channel whose reference is being conditioned.

It is worth mentioning that the objective of the method is neither to shorten the settling time nor to reduce *ISE*. Anyway, by means of an analysis like the one given in Figure 6.16, the designer can know a priori how the method affects other design indexes as the interaction bounds are harder.

6.6   Delay example: catalytic reactor

This example illustrates that the SMRC approach for decentralised control is potentially applicable to dead-time processes, provided a Smith-predictor or similar can be constructed. Indeed, the method implementation does not imply any further complexity for the great number of controlled dead-time processes, which have a predictor already incorporated.

The parameters of the MIMO system considered here have been taken from a pilot plant designed to perform catalytic conversion studies on light petroleum distillation [100]. The plant is basically an adiabatic tubular reactor surrounded by independent electric furnaces. Each one of these heating stages is monitored by a thermocouple. A basic scheme of the reactor is shown in Figure 6.17.

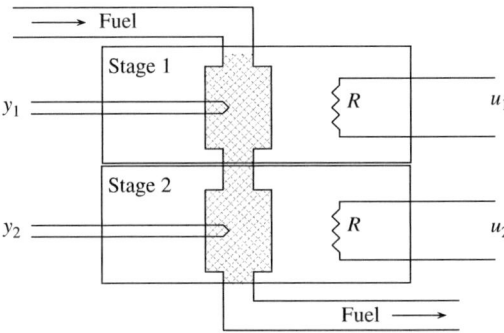

Figure 6.17   Schematic diagram of the catalytic reactor

The system behaviour can be approximately described by the following linearised model:

$$\frac{dx(t)}{dt} = 10^{-4}\begin{bmatrix} -0.51 & 0.21 \\ 0.21 & -0.51 \end{bmatrix} x(t) + 10^{-4}\begin{bmatrix} 0.42 & 0 \\ 0 & 0.42 \end{bmatrix} u(t-490)$$

$$y(t) = \begin{bmatrix} 0.56 & 0.22 \\ 0.22 & 0.56 \end{bmatrix} x(t)$$

(6.58)

where output $y(t) = [y_1 \; y_2]^{\mathrm{T}}$ is a vector with the reactor's inner temperatures and the state $x(t) = [x_1 \; x_2]^{\mathrm{T}}$ is given by the temperatures of the furnace shell. The control action $u(t) = [u_1 \; u_2]^{\mathrm{T}}$ (heating power) has a delay of 490 s.

This plant has the following RGA

$$\Lambda = \begin{bmatrix} 0.6452 & 0.4468 \\ 0.4468 & 0.6452 \end{bmatrix}$$

(6.59)

and then the pairing problem does not demand output inversion. Once more, we add to the system two identical first-order filters $f_1(s)$ and $f_2(s)$, now with eigenvalues

$\lambda_{f_1} = \lambda_{f_2} = -50$. The considered closed loop has already incorporated a predictor and two identical decentralised PI controllers as follows:

$$C_i(s) = 142.3\left(1 + \frac{1}{17,000s}\right), \quad i = 1,2 \tag{6.60}$$

The response of the system when it is excited with two consecutive reference steps is depicted with dashed line in Figure 6.18 (reference signals r_1 and r_2 are drawn with dotted lines). Both the crossed coupling and the dead time of the system can be appreciated.

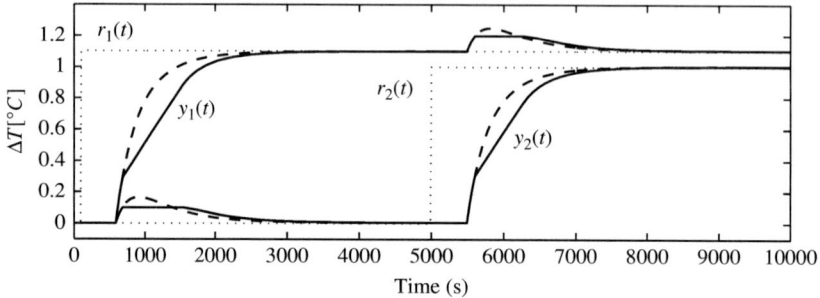

Figure 6.18   Delayed process outputs y with (solid line) and without (dashed line) the conditioning technique

The SMRC algorithm is then implemented to delimit the interaction amplitude to $0.1°C$. Note that the off-diagonal transfer functions of the reactor have the same relative degree as the ones of the inverted quadruple-tank system. Therefore, sliding surfaces of the form of (6.56) and (6.57) will also work for this system. Now we choose constant gains $k_2^1 = 0.1$ and $k_2^2 = 0.1$ for a time constant of approximately $0.1\,s$ in the conditioning loop, whereas we take again $w_1^- = w_2^- = -1$ and $w_1^+ = w_2^+ = 1$.

The solid lines given in Figure 6.18 describe the results obtained. The plain section of the responses corroborate that, even in a dead-time process, interactions are limited to the predefined limits.

Chapter 7

Partial decoupling and non-minimum phase systems

Triangular decoupling has been considered in several contributions as an alternative approach to relax the *dynamic restrictions* imposed by diagonal decoupling in non-minimum phase (NMP) multivariable systems. This strategy allows decoupling a given variable without transferring to its response NMP characteristics. Additionally, since it is based on a centralised architecture, the achievable closed-loop performance is clearly superior to that of decentralised control, particularly for NMP plants.

However, when right-half plane (RHP) zeros are aligned with the variable of interest (the decoupled variable), the aforementioned advantages of partial decoupling come at the cost of large interactions in the remaining system variables. In this chapter, we first study the reasons of this peculiar feature of partially decoupled NMP systems and then employ sliding mode reference conditioning (SMRC) ideas to delimit the remaining interactions without producing inverse responses in the main variable.

7.1 Some introductory comments

We have already seen that crossed interactions among different inputs and outputs are inherent to multivariable systems. In the previous chapter, we discussed an algorithm to delimit interactions in decentralised control structures. However, despite its clear supremacy in process industry, decentralised control does not always lead to adequate solutions. In fact, NMP systems often require centralised control architectures in order to satisfy their performance specifications (see conclusions of Section 6.2).

Ideally, with the aim of cancelling crossed interactions in NMP systems, one could think about a centralised design that achieves full dynamic decoupling of the system, i.e. to obtain a diagonal closed-loop transfer matrix. This could be attained by following, for instance, the procedure presented in Chapter 4 (Section 4.3.2) to synthesise stable approximate inverses of NMP multivariable systems. However, in spite of its intuitive advantages, full decoupling has also an associated performance cost. As discussed in Chapter 4, when the system is NMP diagonal decoupling introduces additional RHP zeros and makes fundamental design limitations of feedback systems harder [11,22,110].

For these reasons, the so-called triangular or partial decoupling has been considered as an alternative way of reducing interactions, as it softens the costs of full dynamic decoupling [47,73,74]. A partially decoupled system has at least one decoupled variable, and its resultant closed-loop transfer matrix is usually lower or upper triangular. Although any system that may be diagonally decoupled may also be put into triangular form, the converse statement is not true [89,143]. This means that partial decoupling is less restrictive than diagonal decoupling, and thus, it is applicable to a greater number of cases. Furthermore, the non-zero off-diagonal elements in the closed-loop transfer matrix help to relax the bounds on sensitivity functions imposed by the RHP zeros in MIMO control systems [110]. Another interesting feature of partial decoupling is that it permits pushing the effects of an NMP zero to a particular output, depending on the zero direction. The price to be paid when exploiting partial decoupling is the presence of interactions in a structured form. As we will see, these interactions strongly depend on the RHP zero direction.

In the following, we first explore the design of centralised controllers that decouple a given variable of NMP processes. Both RHP zero direction and relative degree considerations are combined in order to achieve stable and proper controllers leading to triangular complementary sensitivities. Then, we describe two methodologies that allow limiting the remaining interactions in partially decoupled systems: a simple method that interpolates diagonal and partial decoupling controllers [147], and a supervisory SMRC algorithm based on the main ideas of this book, which avoids as long as possible inverse responses in the decoupled variable.

7.2   Right-half plane zeros directionality and partial decoupling

We will now see what can be gained from relaxing full decoupling requirement. The subsequent results are based on Assumption 7.1.

Assumption 7.1. *(a) All the RHP zeros z_0 of $P(s) \in \mathbb{R}^{m \times m}(s)$ have unitary geometric multiplicity. That is, the rank of $P(z_0)$ is $(m-1)$. (b) There are no poles at $s = z_0$.*

7.2.1   *Algebraic interpolation constraint*

First, we illustrate the closed-loop constraints of triangular decoupling through the plant (4.99) discussed in Example 4.7.

Example 7.1: Let us consider again the NMP system

$$P(s) = \frac{1}{(s+1)(s+3)} \begin{bmatrix} s+3 & 4 \\ 2 & 2 \end{bmatrix} \tag{7.1}$$

We assume now that partial decoupling is aimed, rather than the diagonal decoupling sought for in Example 4.7. Without loss of generality, we take the case in which the desired complementary sensitivity is an upper triangular transfer matrix.

$$T_D(s) = P(s)Q(s) = \begin{bmatrix} t_{11}(s) & t_{12}(s) \\ 0 & t_{22}(s) \end{bmatrix} \tag{7.2}$$

The IMC controller should then have the form

$$Q(s) = P^{-1}(s) \begin{bmatrix} t_{11}(s) & t_{12}(s) \\ 0 & t_{22}(s) \end{bmatrix} \tag{7.3}$$

where the plant inverse is given by (4.101). So,

$$P^{-1}(s) = \frac{(s+1)(s+3)}{2(s-1)} \begin{bmatrix} 2 & -4 \\ -2 & s+3 \end{bmatrix} \tag{7.4}$$

and

$$Q(s) = \frac{(s+1)(s+3)}{2(s-1)} \begin{bmatrix} 2t_{11}(s) & 2t_{12}(s) - 4t_{22}(s) \\ -2t_{11}(s) & -2t_{12}(s) + (s+3)t_{22}(s) \end{bmatrix} \tag{7.5}$$

This implies that for $Q(s)$ to be stable (and consequently for the feedback system to be internally stable), the following conditions must hold:

$$t_{11}(s)|_{s=1} = 0 \tag{7.6}$$

$$2t_{12}(s) - 4t_{22}(s)|_{s=1} = 0 \tag{7.7}$$

Consequently, although the RHP zero is present in the first channel ($t_{11}(s)$), by allowing some degree of one-way coupling one can avoid the presence of the RHP zero in $t_{22}(s)$ (which had caused diagonal decoupling discussed in Example 4.7). What must be guaranteed for the closed-loop internal stability is that $t_{22}(1) = t_{12}(1)/2$.

Obviously, if we had taken a lower-triangular complementary sensitivity, the RHP zero would have appeared in $t_{22}(s)$ but not necessarily in $t_{11}(s)$.

The result obtained in the previous example is a particular case of Theorem 7.1 [88], which provides a direct way to verify when a stable controller $Q(s)$ exists for a given complementary sensitivity.

Theorem 7.1: *There exists a stable controller $Q(s)$ such that the complementary sensitivity is equal to a desired transfer matrix $T_D(s)$ if and only if $T_D(s)$ satisfies*

$$h^T T_D(z_0) = 0 \in \mathbb{R}^m \tag{7.8}$$

for all the RHP zeros z_0 of the plant $P(s)$, with h^T being the output direction of the zero z_0. Condition (7.8) is known as algebraic interpolation constraint.

Proof:

$\Rightarrow$ Assuming there exists a stable $Q(s)$ such that $T_D(s) = P(s)Q(s)$, then it verifies $h^T T_D(z_0) = h^T P(z_0)Q(z_0) = 0 \in \mathbb{R}^m$.

$\Leftarrow$ Consider the following partial fraction expansion of $P^{-1}(s)$

$$P^{-1}(s) = \frac{1}{s - z_0} R_0 + P_z(s) \tag{7.9}$$

where R_0 is the matrix of residues and $P_z(s)$ is a remaining term without poles at $s = z_0$. Post-multiplying both sides of (7.9) by $P(s)$, we get

$$I = \frac{1}{s - z_0} R_0 P(s) + P_z(s)P(s) \tag{7.10}$$

As the left hand side of (7.10) is the identity matrix, the transfer matrix of the right-hand side cannot have poles at $s = z_0$. Note that $P_z(s)P(s)$ has no poles at $s = z_0$ because of Assumption 7.1, and therefore the following must hold:

$$R_0 P(z_0) = 0 \in \mathbb{R}^{m \times m} \tag{7.11}$$

Because z_0 has unitary geometric multiplicity, $P(z_0)$ has rank $(m - 1)$ and R_0 has rank equal to 1. Then, according to the definition of multivariable zeros and its output directions, the rows of R_0 will be multiples of the output direction h^T of the RHP zero z_0. Thus, $h^T T_D(z_0) = 0$ implies that

$$R_0 T_D(z_0) = 0 \in \mathbb{R}^{m \times m} \tag{7.12}$$

If we now post-multiply both sides of (7.9) by $T_D(s)$, we have

$$Q(s) = P^{-1}(s)T_D(s) = \frac{1}{s - z_0} R_0 T_D(s) + P_z(s)T_D(s) \tag{7.13}$$

Since $P_z(s)T_D(s)$ has no poles at $s = z_0$, from (7.12) it is concluded that $Q(s)$ does not have poles at $s = z_0$ either.

Remark 7.1: *From Theorem 7.1 it is straightforward that for Example 7.1*

$$h^T T_D(1) = [1 - 2] \begin{bmatrix} t_{11}(1) & t_{12}(1) \\ 0 & t_{22}(1) \end{bmatrix} = [t_{11}(1) \quad t_{12}(1) - 2t_{22}(1)] = 0 \in \mathbb{R}^2 \tag{7.14}$$

which coincides with the necessary and sufficient conditions (7.6) and (7.7) for the closed loop to be internally stable.

The interpolation constraint (7.8) requires that each column of the desired closed-loop transfer matrix $T_D(s)$ is orthogonal to h^T when evaluated at the plant zero z_0.

Another consequence of (7.8) is that RHP zeros z_0 have no effect on the outputs corresponding to zero elements of h^T (as happens with plant $P'(s)$ in Example 4.7).

Now, if full decoupling is aimed ($T_D(s)$ diagonal), the interpolation constraint (7.8) imposes that the degree of z_0 in $T_D(s)$ has to be at least equal to the number of non-zero entries in h. Hence, for the general case in which h has more than one non-zero entry, this is another way to explain why full dynamic decoupling spreads the RHP zeros among the corresponding channels of the closed loop (see Section 4.4).

In contrast, partial decoupling allows pushing the effect of the RHP zero to a particular controlled variable, which may be less important to control tightly. This is possible because, although the interpolation constraint requires certain relationship between the elements of each column of $T_D(s)$, the columns themselves can be chosen independently.

7.2.2   Inverse response on a particular output

Theorem 7.2 reveals the directional effect of RHP zeros in partially decoupled systems.

Theorem 7.2: *Assume $P(s)$ has a unique RHP zero z_0, and that the kth element of its output direction h_k is non-zero. Then, to obtain an internally stable closed-loop system, the desired complementary sensitivity can be chosen as follows:*

$$T_D(s) = \begin{bmatrix} 1 & 0 & \cdots & 0 & 0 & 0 & \cdots & 0 \\ 0 & 1 & \cdots & 0 & 0 & 0 & \cdots & 0 \\ \vdots & \vdots & \ddots & \vdots & \vdots & \vdots & \cdots & \vdots \\ \vdots & \vdots & & 1 & \vdots & \vdots & \cdots & \vdots \\ \dfrac{\beta_1 s}{s+z_0} & \dfrac{\beta_2 s}{s+z_0} & \cdots & \dfrac{\beta_{k-1} s}{s+z_0} & \dfrac{-s+z_0}{s+z_0} & \dfrac{\beta_{k+1} s}{s+z_0} & \cdots & \dfrac{\beta_m s}{s+z_0} \\ \vdots & \vdots & \cdots & \vdots & \vdots & 1 & \cdots & \vdots \\ \vdots & \vdots & \cdots & \vdots & \vdots & \vdots & \ddots & \vdots \\ 0 & 0 & \cdots & 0 & 0 & \cdots & 0 & 1 \end{bmatrix}$$

$$(7.15)$$

where

$$\beta_j = -\frac{2h_j}{h_k}, \quad j \neq k \tag{7.16}$$

Proof: Trivial from the interpolation constraint (7.8).

This result quantifies the effect of moving completely an RHP zero to a given output y_k, owing to the zero direction. As can be observed, if the zero is naturally aligned with this output ($h_k \gg h_j$), only small interactions will occur as a consequence of the zero pushing. Conversely, if the zero is predominantly aligned with a decoupled output y_j, $j \neq k$, pushing the zero to output y_k will produce large interactions ($\beta_j \gg 1$). In any case, one can always achieve static decoupling (i.e. suppression of steady-state interactions) by means of the zero at the origin in the off-diagonal elements of (7.15).

Although Theorem 7.2 gives an insight about the directional effect of RHP zeros in partial decoupling, it assumes perfect control to be reachable and thus it does not guarantee the existence of a proper $Q(s)$. To this end, apart from satisfying the interpolation constraint (7.8), the entries of the complementary sensitivity matrix must be chosen with the appropriate relative degree. In Reference 49, a methodology for the synthesis of stable and proper $Q(s)$ is presented, based on the factorisation of NMP plants enabled by the so-called z-interactors (see Remark 4.4). The latter describes the RHP finite zero structure of a system in a similar way as the interactors studied in Chapter 4 do for the relative degree structure (infinite zeros). Anyway, one must finally choose $T_D(s)$ (or a factor of it) so as to make $Q(s) = P^{-1}(s)T_D(s)$ proper. Note in particular that if a biproper $Q(s)$ is aimed, $T_D^{-1}(s)$ must have the same relative degree structure as the left interactor $\xi_l(s)$ of $P(s)$.

As a consequence of these relative degree requirements, the magnitude of interactions depends not only on the zero direction but also on the zero location and the relative degrees of the transfer functions that compose $T_D(s)$. However, the way in which the zero direction affects interactions is still the one described by Theorem 7.2. Example 7.2 illustrates both statements.

Example 7.2: Consider again the plant model (5.24) discussed in Section 5.5.1

$$P(s) = \frac{1}{(s+1)^2} \begin{bmatrix} s+2 & -3 \\ -2 & 1 \end{bmatrix} \tag{7.17}$$

where $P(s)$ is an NMP plant with a transmission zero at $s = z_0 = 4$. The zero output direction is $h^T = [1\ 3]$, which explains the NMP behaviour obtained in y_2 when the system was fully decoupled in Chapter 5 (see Figure 5.7). Taking into account the relative degrees in (7.17), the zero direction of z_0 and the interpolation constraint (7.8), we propose upper and lower one-way decoupled complementary sensitivities for which stable and proper $Q(s)$'s exist.

$$T_{lo}(s) = \begin{bmatrix} \dfrac{4}{s+4} & 0 \\[3mm] \dfrac{\alpha_1 s}{(s+4)^3} & \dfrac{16(-s+4)}{(s+4)^3} \end{bmatrix}$$

$$T_{up}(s) = \begin{bmatrix} \dfrac{4(-s+4)}{(s+4)^2} & \dfrac{\alpha_2 s}{(s+4)^2} \\[3mm] 0 & \dfrac{16}{(s+4)^2} \end{bmatrix} \tag{7.18}$$

It can be easily verified that both $T_{up}(s)$ and $T_{lo}(s)$ make $Q(s)$ proper. The parameters α_1 and α_2 have to be determined in order to satisfy (7.8). Hence,

$$[1\ 3]T_{lo}(4) = 0 \Rightarrow \alpha_1 = \frac{-64}{3}$$
$$[1\ 3]T_{up}(4) = 0 \Rightarrow \alpha_2 = -12$$

Note that the differences between the α values and the β values (that would have been obtained from h^{T} according to Theorem 7.2) are related to the different relative degree of the off-diagonal elements in $T_{up}(s)$ and $T_{lo}(s)$. However, the directional effect of z_0 captured by Theorem 7.2 is still present in this example, as can be appreciated in the simulations shown in Figure 7.1. Therein, it is evidenced that completely moving the effect of z_0 to output y_1 produces large interactions (top plot of the figure), while smaller coupling arises from shifting the zero to output y_2 (bottom box). This is consistent with the previous analysis, since z_0 is mainly aligned with output y_2.

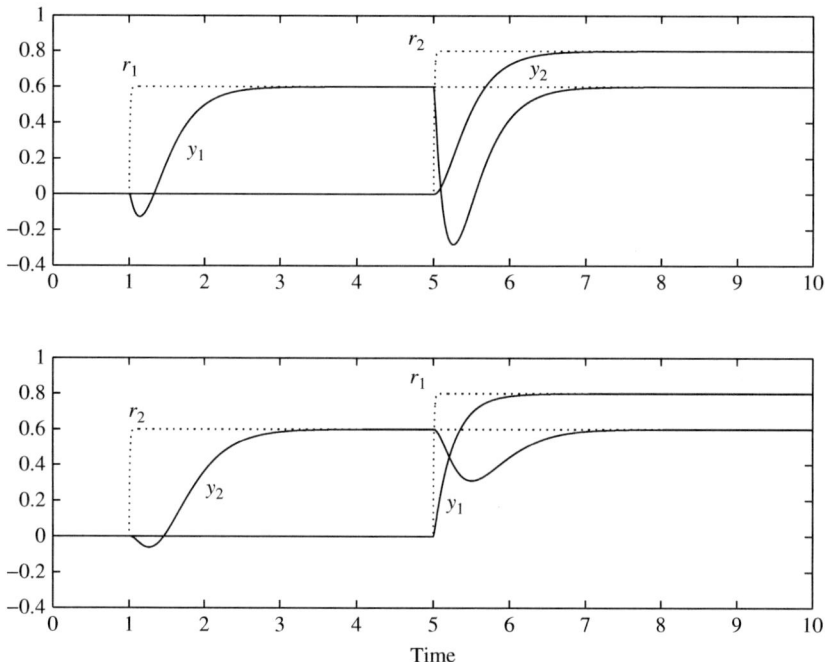

Figure 7.1   Upper (top box) and lower (bottom box) triangular decoupling of the plant represented by (7.17)

From the reasoning and aforementioned example, one could think of using RHP zero directions and relative degrees as a guide to minimise interactions in one-way decoupling designs. However, triangular decoupling is more likely to be used to control processes in which a given variable should be controlled tighter than others. Therefore, the variable of interest cannot be chosen arbitrarily, but it generally depends on the nature of the process. If because of the structure of the plant the RHP zero is mainly aligned with the variable of interest, the previous result states that pushing the zero to another output will come at the cost of great interactions. In the following sections, we discuss a couple of methods to limit these interactions.

7.3   Interpolating diagonal and partial decoupling

A method that allows the designer to trade off partial decoupling advantages against interaction tolerance was presented by Weller and Goodwin [146,147], where a single scalar parameter is used to interpolate diagonal and partial decoupling controllers.

Based on the interactor matrices studied in Chapter 4, they proposed a couple of IMC controllers $Q_{PD}(s)$ and $Q_{DD}(s)$ that achieve partial and diagonal decoupling of NMP plants, respectively. Then, they introduced a scalar tuning variable λ to blend the designs for partial and diagonal decoupling as follows:

$$Q(s) = \lambda Q_{DD}(s) + (1 - \lambda)Q_{PD}(s), \quad 0 \le \lambda \le 1 \qquad (7.19)$$

thereby providing freedom to address the compromise between transient cross coupling magnitude and direct closed-loop response on the main (decoupled) variable.

Although this method is effective and provides easy tunings, for NMP systems it gives rise to undershoots in the decoupled variable as interactions are made smaller. This is not a minor issue, since, as already argued, a tight control of this variable is usually required in partially decoupled designs. The amplitude of the resulting undershoots can be quantified by the bounds derived in Reference 67, where it is shown that RHP zeros close to the origin and small interactions produce large undershoots in step responses.

Both the advantages and disadvantages of this simple method are illustrated in Section 7.6, where we evaluate its capability of reducing interactions in the quadruple-tank process when subjected to demanding control requirements.

7.4   Partial decoupling with bounded interactions via SMRC

We present here an alternative methodology to enhance partially decoupled NMP systems [39]. It aims at limiting the remaining coupling without transferring NMP dynamics to the main (decoupled) variable. To this end, the reference signal for the main variable is conditioned in the same line as the ideas presented in Chapter 6 for decentralised control structures, i.e. via an SMRC loop. Although the methodology presented in Chapter 6 notoriously improves the performance of pre-existing multiloop control systems, it also inherits the inherent limitations of decentralised control. For instance, it was shown for the quadruple-tank process that trying to eliminate interactions on at least one system variable led to a serious degradation of the closed-loop performance. The use of a centralised control strategy allows, apart from a greater degree of decoupling, achieving a closed-loop performance highly superior to the one obtained in the previous chapter. These advantages of partial decoupling will be verified for the quadruple-tank process, using both the Weller–Goodwin method just described and the SMRC approach to be discussed below.

Figure 7.2 illustrates a standard 2×2 partially decoupled system (inside the dotted box) to which an outer conditioning loop was added for interaction limitation. $P(s)$ represents the process to be controlled and $C(s)$ is a controller that achieves upper-triangular decoupling, configuration to be considered in this subsection without loosing generality. For lower-triangular decoupling, one should simply mirror references and boundary subindexes.

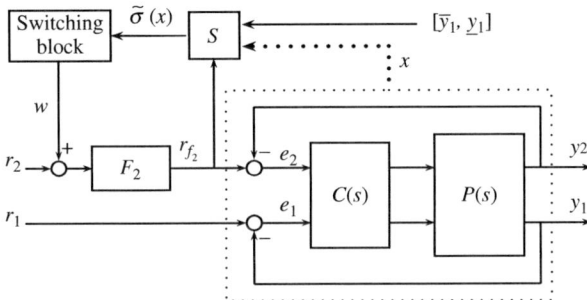

Figure 7.2 Partially decoupled system ($T(s)$ upper triangular) with outer reference conditioning loop

As can be seen, the structure of the conditioning loop is quite similar to the one described in Chapter 6, so it will be only briefly analysed. Instead, we will focus on the different features of the present solution (addressed problem, required conditions for method application, reachable performance, etc.) and on its comparison with the alternative approach discussed in Section 7.3.

Block S computes switching functions $\underline{\sigma}(x)$ and $\bar{\sigma}(x)$ taking into account the lower and the upper interaction limits $\underline{y}_1$ and $\bar{y}_1$, respectively (again, $\tilde{\sigma}(x)$ stands for either $\underline{\sigma}(x)$ or $\bar{\sigma}(x)$). $F_2(s)$ is the first-order linear filter usually included in SMRC schemes, dynamics of which is chosen faster than the closed-loop one.

The commutation law implemented in the *switching block* is then

$$\begin{cases} w = w^- & \text{if} \quad \bar{\sigma}(x) < 0 \\ w = w^+ & \text{if} \quad \underline{\sigma}(x) > 0 \\ w = 0 & \text{otherwise} \end{cases} \tag{7.20}$$

with

$$\bar{\sigma}(x) = \tilde{y}_1 - y_1 - \sum_{\alpha=1}^{\rho} k_{\alpha+1} y_1^{(\alpha)} \tag{7.21}$$

where ρ is the relative degree of the transfer function between output y_1 and reference r_{f_2}, $y_1^{(\alpha)}$ is the αth derivative of y_1 and $k_{\alpha+1}$ are constant gains.[1]

As we have already seen, the switching logic ((7.20) and (7.21)) leads the filtered reference r_{f_2} to be transiently conditioned when there is a risk of exceeding the established coupling limits. In the present application, this reference dosage allows limiting the remaining crossed coupling without producing inverse responses in the decoupled variable.

When the SMRC approach is applied to a problem where only y_1 regulation is aimed, the limits $\overline{y_1}$ and $\underline{y_1}$ are constant values that determine security bounds on the gap between the variable y_1 and its set point r_1. However, in those cases where a simultaneous tracking of r_1 and r_2 is also an objective, the coupling bounds should be generated as $\overline{y_1} = \hat{y}_1 + \delta_1^+$ and $\underline{y_1} = \hat{y}_1 - \delta_1^-$, with $\hat{y}_1$ being the predicted evolution of y_1 due to a change exclusively in r_1 (as if the system was diagonally decoupled), and $\delta_1^\pm$ constant interaction tolerances. Note that $\hat{y}_1$ can be easily obtained from r_1 changes provided the plant model is known. Besides, model uncertainties could be taken into account by a conservative choice of $\delta_1^\pm$. Therefore, if both reference signals changed together, the outer conditioning of r_{f_2} would become active only if the response of y_1 was about to differ from the (artificially generated) decoupled response $\hat{y}_1$ more than what is permitted by the tolerances $\delta_1^\pm$. This operation is illustrated in Section 7.5.[2]

Observe that in the present application the hidden dynamics during SMRC is determined by the zeros of the closed-loop transfer function between y_1 and r_{f_2}. Since the off-diagonal transfer functions of a partially decoupled control system can always be designed to have all their zeros in the LHP (as explained in Section 7.2), the hidden dynamics will be stable. Furthermore, the SMRC will conclude when the output of the system returns by itself to the interior of the region between the coupling limits, which always occurs because the main control design achieves static decoupling (see $T(s)$ in Theorem 7.2), and thus there are no interactions in steady state.

7.5 Numerical example

Consider again Example 7.2 discussed in Section 7.2.2 in which the process is described by (7.17). The closed-loop transfer matrix $T_{up}(s)$ in (7.18) is obtained by means of the IMC controller:

$$Q_{up}(s) = \frac{(s+1)^2}{(s+4)^2} \begin{bmatrix} -4 & -12 \\ -8 & -8 \end{bmatrix} \qquad (7.22)$$

---

[1] Recall that $y_1^{(\alpha)}$ is not computed by differentiating y_1, but it is generated as a linear combination of the closed-loop states by means of a linear transformation like the one detailed in the previous chapters.

[2] The same prediction of the ideal decoupled output could be applied for the decentralised control structure of the previous chapter in order to relax the assumption of no simultaneous reference changes.

Because of the RHP zero direction, y_2 decoupling causes great interactions on y_1 (see Figure 7.1). The objective is then to confine these interactions within pre-defined bounds without degrading considerably the response of y_2. Hence, the SMRC method was added as shown in Figure 7.2. Since the off-diagonal element of $T_{up}(s)$ has unitary relative degree ($\rho = 1$), the following choice was made for the sliding surfaces:

$$\tilde{\sigma}(x) = \tilde{y}_1 - y_1 - k_2 \dot{y}_1 = 0 \tag{7.23}$$

The eigenvalue of filter $F_2(s)$ was set at $\lambda_f = -100$ and the values of the discontinuous signal were $w^+ = -w^- = 1$.

To illustrate the generation of the output derivatives from the system states, the following minimal realisation of (7.17) is considered:

$$\dot{x} = \begin{bmatrix} -2 & 0.5 & 0 & 0 \\ 2 & 0 & 0 & 0 \\ 0 & 0 & -2 & -0.5 \\ 0 & 0 & 2 & 0 \end{bmatrix} x + \begin{bmatrix} 1 & 0 \\ 0 & 0 \\ 0 & 1 \\ 0 & 0 \end{bmatrix} u \tag{7.24}$$

$$y = \begin{bmatrix} 1 & 0 & 0 & -1.5 \\ 0 & 1 & 0 & 0.5 \end{bmatrix} x$$

In this way, sliding surfaces (7.23) were actually implemented as

$$\tilde{\sigma} = \tilde{y}_1 - y_1 - k_2(0.5x_2 - 3x_3 + u_1) = 0 \tag{7.25}$$

Figure 7.3 presents simulation results for $k_2 = 1/100$, which sets a conditioning time constant of approximately 0.01 [*time units*]. The following interaction bounds were considered: $\underline{y_1} = -0.3$ (dotted line), $\underline{y_1} = -0.1$ (dashed-dotted line), $\underline{y_1} = 0.1$ (dashed line) and $\overline{y_1} = 0.3$ (solid line). The corresponding evolution of the other signals in the loop is plotted in Figure 7.4. As can be appreciated, the method successfully limits interaction without affecting the decoupled variable y_2 with inverse responses. Moreover, the original control system is not altered at all when no interaction limit is reached (in the example, for $y_1 \leq -0.3$).

Figure 7.5 illustrates how the rate of approach of the interaction to its limit can be controlled by means of the sliding surface design. Particularly, the limit $\underline{y_1} = 0.1$ of Figure 7.3 was considered, but taking $k_2 = 1/100$ (solid line), $k_2 = 1/20$ (dashed line) and $k_2 = 1/10$ (dotted line).

Finally, the method of operation when both references change together (at $t = 1$) is shown in Figure 7.6. In this case, coupling bounds $\underline{y_1}$ and $\overline{y_1}$ are specified with respect to the decoupled response $\hat{y}_1$ instead of being constant values. Simulations were run for $\delta_1^\pm = 0.3$, $\delta_1^\pm = 0.5$ and $\delta_1^\pm$ large enough to obtain the original closed-loop response (SMRC does not establish in this latter case).

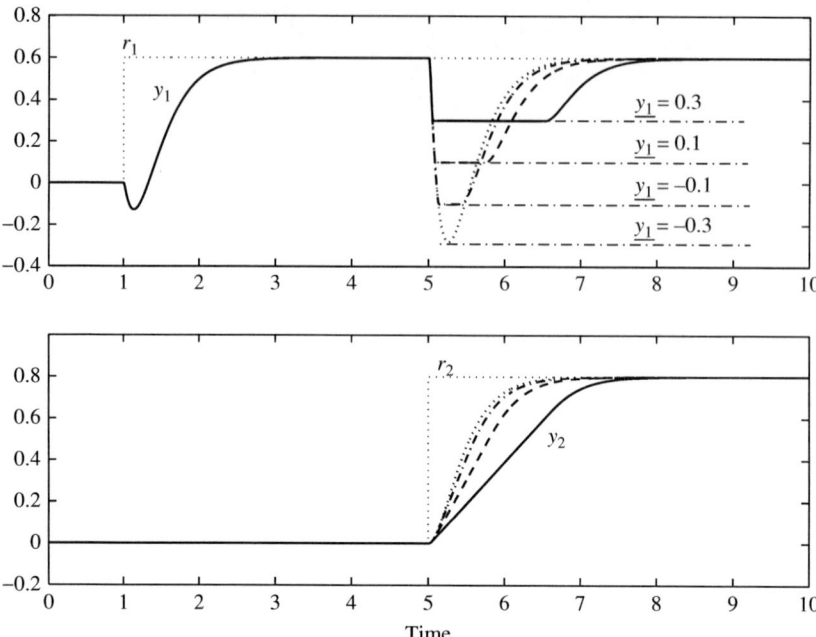

Figure 7.3   Output signals with different interaction bounds

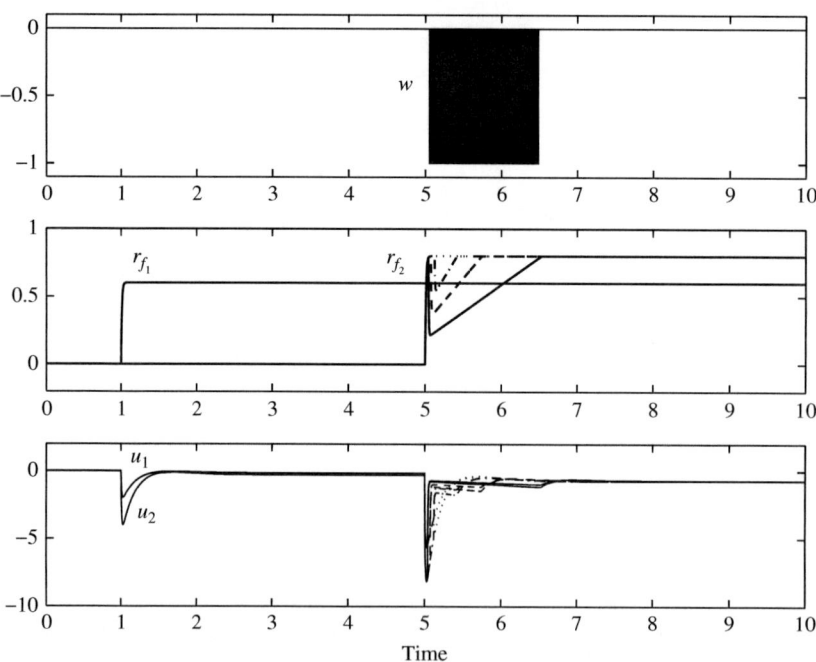

Figure 7.4   Discontinuous signals (overlapped), filtered references and controller outputs u corresponding to Figure 7.3

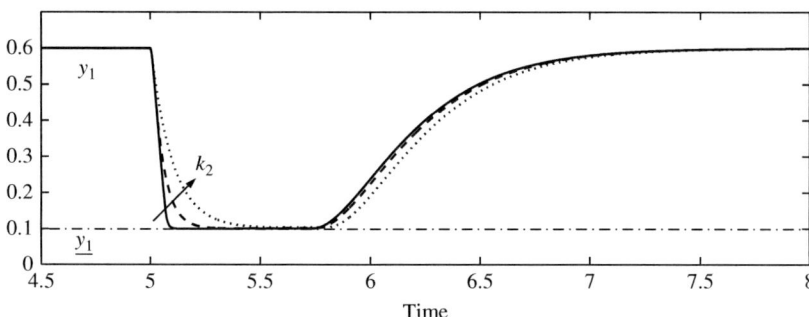

Figure 7.5　Different rates of approach to interaction limits on y_1

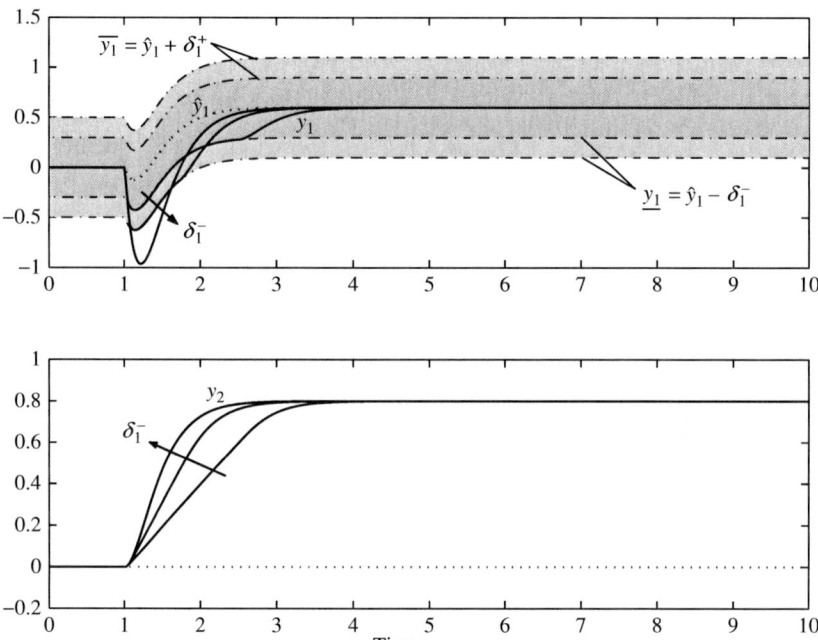

Figure 7.6　Coupling limitation on y_1 during simultaneous tracking of r_1 and r_2, with different interaction tolerances $\delta_1^{\pm}$

7.6　Case study: quadruple tank

We revisit now the quadruple-tank process with an adjustable zero introduced in Chapter 6 [66]. The schematic diagram of this plant was presented in Figure 6.9.

For the transfer matrix $P(s)$ corresponding to the linearised model of system (6.47), the output direction $h_o^T = [h_{o_1} \ h_{o_2}]$ of the zero at $s = z_0 > 0$ is given by[3]

$$
[h_{o_1} \ h_{o_2}]
\begin{bmatrix}
\dfrac{\gamma_1 T_1 k_1 k_c}{A_1(1 + T_1 z_0)} & \dfrac{(1 - \gamma_2) T_1 k_1 k_c}{A_1(1 + T_1 z_0)(1 + T_3 z_0)} \\[4mm]
\dfrac{(1 - \gamma_1) T_2 k_2 k_c}{A_2(1 + T_2 z_0)(1 + T_4 z_0)} & \dfrac{\gamma_2 T_2 k_2 k_c}{A_2(1 + T_2 z_0)}
\end{bmatrix}
= 0 \in \mathbb{R}^2
$$

$$(7.26)$$

It is worth observing from (7.26) that $h_{o_1} \neq 0$ and $h_{o_2} \neq 0$. Thus, the zero will never be aligned with a single input. Indeed, from the first component of (7.26), we get

$$
\frac{h_{o_1}}{h_{o_2}} = -\frac{1 - \gamma_1}{\gamma_1} \cdot \frac{T_2 k_2 A_1(1 + T_1 z_0)}{T_1 k_1 A_2(1 + T_2 z_0)(1 + T_4 z_0)}
$$

$$(7.27)$$

In this way, when the left flow-divider valve is such that γ_1 is close to zero, the multivariable zero is mainly aligned with the first output y_1. Conversely, if γ_1 is close to one, then the zero is mainly associated with the second output y_2. Therefore, the flow-divider valves' settings γ_1 and γ_2 affect, for a given RHP zero location, the alignment of this zero with each controlled variable.

For the operating point $P+$ (see Table 6.1), the transfer matrix (6.47) resulted

$$
P_+(s) =
\begin{bmatrix}
\dfrac{1.5}{1 + 63s} & \dfrac{2.5}{(1 + 39s)(1 + 63s)} \\[4mm]
\dfrac{2.5}{(1 + 56s)(1 + 91s)} & \dfrac{1.6}{1 + 91s}
\end{bmatrix}
$$

$$(7.28)$$

which has a transmission zero at $s = z_0 = 0.013$ with output direction $h_o^T = [h_{o_1} \ h_{o_2}] = [0.626 \ -0.779]$.

It is assumed now that the level in the left-lower tank h_1 is the most important variable and that a tight control of this level is required. Moreover, no greater interactions than the magnitude of the changes in h_1 are tolerated in the secondary variable h_2. Therefore, we look for a proper IMC controller $Q(s)$ leading to a lower-triangular complementary sensitivity and satisfying the interpolation constraint (7.8). To accomplish the former objective, the closed-loop complementary sensitivity is chosen as

---

[3] In this section we denote with h_{o_i} the components of the output direction of the zero at z_0, in order to avoid confusion with the tank levels h_i.

$$T_{lo}(s) = \begin{bmatrix} \dfrac{1}{a_{11}s + 1} & 0 \\ \dfrac{as}{(a_{21}s + 1)^2} & \dfrac{(-1/z_0)(s - z_0)}{(a_{22}s + 1)^2} \end{bmatrix} \tag{7.29}$$

In order to satisfy (7.8), α must be

$$\alpha = \frac{-h_{0_1}}{a_{11}z_0 + 1} \frac{(a_{21}z_0 + 1)^2}{h_{0_2}z_0} \tag{7.30}$$

Furthermore, we require a closed-loop settling time not greater than 200 sec in the main system variable (h_1), i.e. approximately five times faster than the response obtained in Chapter 6 with a decentralised control architecture.

According to the desired closed-loop dynamics and the importance of each variable, we take as a first approach $a_{11} = 20$, $a_{21} = 20$ and $a_{22} = 50$. Simulations were run with the resulting design for steps in r_1 at $t = 50$ s and in r_2 at $t = 500$ s. The top plot in Figure 7.7 depicts the closed-loop responses.[4] Although the main variable h_1 is decoupled and presents no undershoots, the level h_2 is altered beyond the required interaction limits by the step in r_1. A trivial way of reducing this

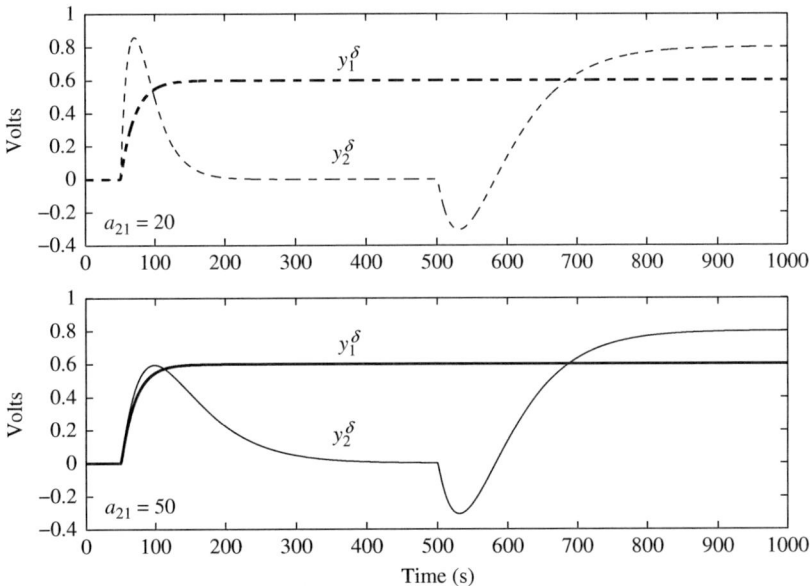

Figure 7.7   Original partially decoupled system. Dashed line: $a_{21} = 20$. Solid line: $a_{21} = 50$

---

[4] Note that we plot here the outputs $y_i^\delta = k_c(h_i - h_i^0)$ with respect to the operating point, differing from Chapter 6 where the system outputs $y_i = k_c h_i$ were shown.

interaction would be to make slower the crossed transfer function of the closed loop. This was performed by choosing $a_{21} = 50$, and the results are shown in the bottom box of Figure 7.7. In addition to the longer transient on y_2^δ exhibited in the figure, the integral error of the interaction enlarges considerably. Indeed, the *integral absolute error (IAE)* of y_2^δ for the step in r_1 (until $t = 500$ s) rises from $IAE_2 = 47$ with $a_{21} = 20$ to $IAE_2 = 77$ for $a_{21} = 50$. Obviously, the same index value is achieved for both cases in the main variable y_1^δ ($IAE_1 = 12.6$).

Second, the Weller–Goodwin method discussed in Section 7.8 was considered to improve the response of the partially decoupled system under output (interaction) bounds. For comparative purposes, we take from now on $a_{21} = 20$. The results obtained with this method are shown with solid lines in the top plot of Figure 7.8. They reveal that the interpolation of diagonal and partial decoupling designs produces a significant improvement of the transient response on y_2^δ against the step in r_1 (the previous design responses with $a_{21} = 20$ are repeated in dashed lines). However, a slower and inverse response appears in the main output y_1^δ. Both phenomena are quantified by the IAE: the great reduction of IAE_2 (from 47 to 32.5) contrasts with a larger increase of IAE_1 (from 12.6 to 30.5).

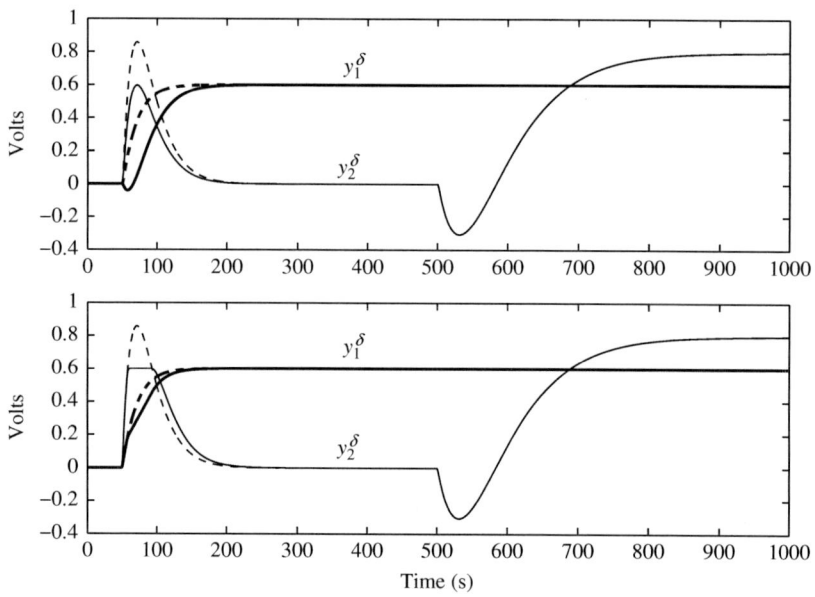

Figure 7.8   Dashed line: Original partially decoupled system ($a_{21} = 20$). Solid line: System with Weller–Goodwin *(top) and SMRC (bottom) methods*

Finally, the SMRC algorithm was tested. To this end, the filter eigenvalue was set as $\lambda_f = -1$, while the bounding surfaces were defined as ($k_2 = 1$) follows:

$$\tilde{s} = \tilde{y}_2^\delta - y_2^\delta - \dot{y}_2^\delta = 0 \tag{7.31}$$

To compute the output derivative we should replace in the state-space representation of the quadruple tank (6.45) the values given in Table 6.1 and the time constants for the operating point P_+, resulting in

$$\frac{dx}{dt} = \begin{bmatrix} -0.0159 & 0 & 0.0256 & 0 \\ 0 & -0.011 & 0 & 0.0179 \\ 0 & 0 & -0.0256 & 0 \\ 0 & 0 & 0 & -0.0179 \end{bmatrix} x + \begin{bmatrix} 0.0482 & 0 \\ 0 & 0.035 \\ 0 & 0.0775 \\ 0.0559 & 0 \end{bmatrix} u$$

$$y^\delta = \begin{bmatrix} 0.5 & 0 & 0 & 0 \\ 0 & 0.5 & 0 & 0 \end{bmatrix} x$$

$$(7.32)$$

Hence, the surfaces (7.31) can be implemented as

$$\tilde{s} = \tilde{y}_2^\delta - y_2^\delta + 0.0055x_2 - 0.0089x_4 - 0.0175u_2 = 0 \tag{7.33}$$

The bottom plot in Figure 7.8 (solid lines) demonstrates that the method effectively confines interaction to the allowed magnitude (amplitude of step in r_1) without transferring NMP characteristics to y_1^δ. Again, the original system responses are drawn in dashed lines. Regarding the integral error, the SMRC strategy achieves the same IAE_2 as the one obtained with the original system, while the measure on the main variable is significantly smaller than with the previous method ($IAE_1 = 17$).

As a concluding remark, it is important to emphasise that in addition to the complete decoupling of the main variable, the use of a centralised strategy allowed achieving responses many times faster than the ones reached with decentralised control. In fact, if for the quadruple-tank process with the multiloop controller discussed in Chapter 6 one tried to approximate the response times shown in Figure 7.8 (bottom), the closed loop would become unstable. Although as already warned there is room for improvement in the design of those PI controllers, the inherent limitations of the control architecture make impossible in general to match the centralised controller performance.

Chapter 8

MIMO bumpless transfer

This chapter describes the undesired effects caused by manual–automatic or controller switching in process control. Manual–automatic commutations are often used in the start-up schemes of the process industry, whereas switching among different controllers is frequently used for the control of non-linear systems at given operating points. Both types of switching at the plant input can be attributed to *structural constraints*, limiting either the controller type (linear controllers are typically preferred) or the switching policy (industry applications seldom include sophisticated approaches, like linear parameter-varying).

The performance degradation caused by these mode switches is first tackled by means of sliding mode reference conditioning (SMRC) ideas in single-input single-output (SISO) systems. Then, multivariable sliding mode (SM) concepts are introduced and the SMRC bumpless algorithm is extended to deal with multiple-input multiple-output (MIMO) processes, so as to avoid inconsistencies between the off-line controller outputs and the plant inputs. As a consequence, jumps at the plant inputs are prevented and undesired transients on controlled variables are significantly reduced.

8.1 Introduction

A common start-up operation in process control, especially in chemical industry, is to take the plant manually to the operating point and then to connect the controller. Another frequent practice is to control non-linear systems at different operating points by switching between various linear controllers [79]. However, such mode switches may cause jumps at the plant inputs and deterioration of the system response if nothing is done to avoid it. The suppression of these jumps at the plant inputs and its associated transient effects is referred to as *bumpless transfer*.

This is a problem of great practical importance for engineers in industry, and consequently there has been a lot of research in this area. Many contributions have dealt with bumpy transfers together with windup problem because of their similar causes. Among the large number of articles that have been subsequently reported in the literature, concepts of 'realisable reference' [55], linear quadratic theory [131], linear matrix inequalities [90], L_2 bounds on state mismatch [155], state/output feedback [160] and H_∞ optimisation [25] have been exploited to find solutions to bumpy transfers.

This chapter first introduces this problem by providing a SISO example already employed to illustrate controller windup in Chapter 1. It then shows a simple and direct way of solving the problem by means of SMRC techniques. Finally, multivariable concepts of VSS and SM are discussed to solve the problems that arise from open-loop to closed-loop (OL–CL) switching or from commutations between different controllers in MIMO systems. The resulting strategy is applicable to controllers for which conventional bumpless algorithms were not conceived, e.g. centralised multivariable controllers, and even then it requires minimal design and implementation effort. Besides, the method also presents robustness properties and does not require a model of the plant.

8.2   Switching at the plant input

Example 8.1 illustrates the effects of plant input switching by means of a simple SISO example.

Example 8.1: Recall Example 1.1 used to show controller windup. Let the plant model and the set PI–Smith predictor be given again by (1.1)–(1.3). Now, a switch at the plant input is considered instead of actuator saturation.

With the purpose of evaluating the consequences of plant input switches, it is assumed here that the PI controller has been designed to operate around the unitary value of the controlled variable y. Hence, the process is driven manually until the output y reaches the 90% of its set point ($r = 1$), and at that moment the feedback loop is closed.

The block diagram of the manual–automatic control system is depicted in Figure 8.1. Therein, the Smith predictor was considered as a built-in part of the process. If this were not the case, the predictor should be implemented within the controller, and therefore the switch should be placed at point A in Figure 8.1. For simulations we have chosen the configuration shown in Figure 8.1, because this is the case in which greater discontinuities occur at the plant input. However, it is worth mentioning that the switch position affects neither the conclusions of this section nor the results of the next one.

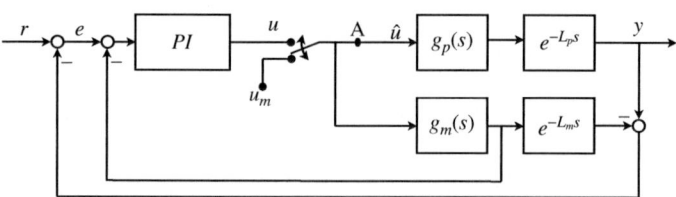

Figure 8.1   Manual–automatic control scheme for the system discussed in Example 1.1

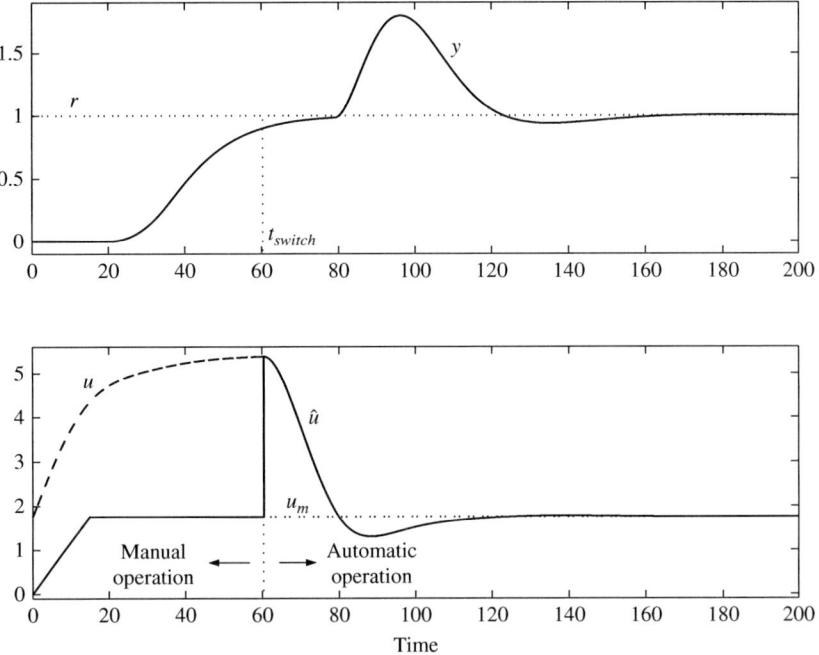

Figure 8.2   Jump at the plant input and transient on controlled variable when switching from manual to automatic mode

---

Figure 8.2 shows the results obtained when the system is switched from manual to automatic mode as mentioned above. As can be appreciated, during the manual operation of the system the controller output (u) becomes much larger than the manually manipulated control action (u_m). This leads to a great jump at the plant input ($\hat{u}$) when the loop is closed at t_{switch}, which in turn produces an undesired response on the controlled variable.

8.3   A simple SMRC solution for SISO systems

The *bumpy transfer* discussed in Example 8.1 can also be tackled using SMRC ideas. Actually, the same switching function and law as in (2.4) and (2.5) can be defined with this purpose assuming biproper controllers. That is

$$
w = \begin{cases} w^- & \text{if} \quad \sigma < 0 \\ w^+ & \text{if} \quad \sigma > 0 \\ 0 & \text{if} \quad \sigma = 0 \end{cases}
\tag{8.1}
$$

where w^-, $w^+ \in \mathbb{R}$, $w^- \neq w^+$ and

$$\sigma = \hat{u} - u \tag{8.2}$$

where $\hat{u}$ is the plant input and u is the controller output. Observe that, unlike most of the SMRC approaches discussed so far, the switching policy (8.1) enforces in this application to reach the surface from both sides (i.e. it works like a conventional SM control), because $\hat{u}$ will be in general different from u during manual mode.

In this way, provided $w^{\pm}$ satisfies the necessary and sufficient condition for SM existence, the trivial switching function (8.2) will make the off-line controller output u to coincide with the manually driven input u_m during manual mode. Consequently, it will avoid jumps at the plant input when the feedback loop is closed.

The discontinuous signal w must naturally be passed through a first-order low-pass filter to smooth out the main loop signals, as was done in the SMRC schemes discussed earlier. Let us see how this SMRC application is implemented in Example 8.1.

Example 8.1[Continued]: To solve the bumpy problem illustrated in Section 8.2, we propose the scheme given in Figure 8.3, where both (8.1) and (8.2) are implemented. An eigenvalue $\lambda_f = -0.5$ was set for the first-order filter $f(s)$, whereas the discontinuous amplitude was taken as $w^{\pm} = \pm 1$.

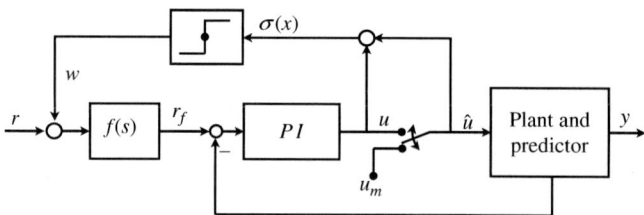

Figure 8.3   SMRC bumpless scheme for the system shown in Figure 8.1

The corresponding results are depicted in Figure 8.4. Clearly, the undesired transient on the controlled variable y is removed. This is attained because the establishment of a sliding regime on the surface $\sigma(x) = \hat{u} - u = 0$ generates the necessary (off-line) reference r_f so that the controller output coincides with the plant input, i.e. $u \equiv \hat{u}$ (see the lower box in Figure 8.4). Hence, the jump at the plant input by the moment of mode switching ($t = t_{switch}$) is avoided. Once the position of the switch is changed, the controller output is directly the plant input, thus $w = 0$ from $t = t_{switch}$ and the SMRC loop becomes inactive. Note also that after commutation to automatic mode the system responds with the closed-loop dynamics to the remaining step of amplitude 0.1.

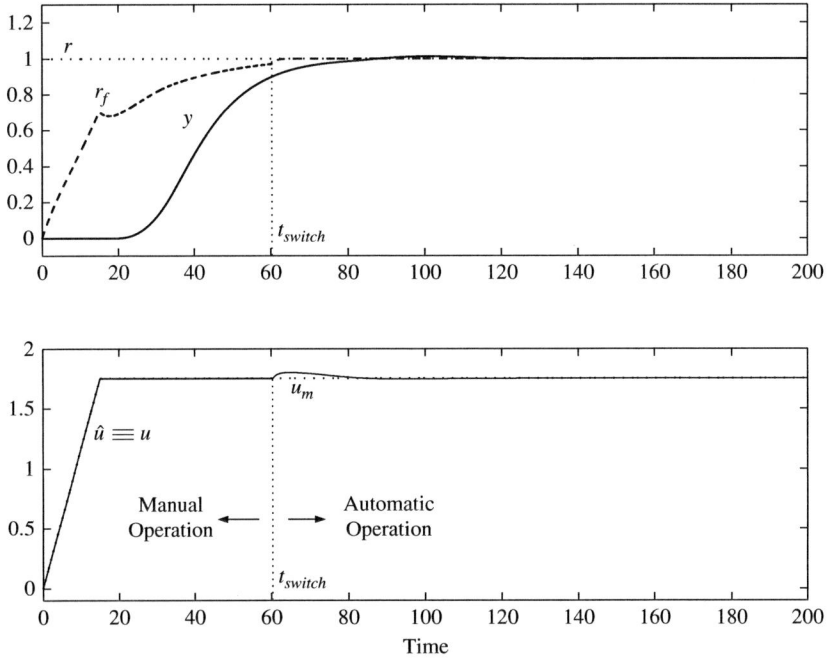

Figure 8.4   Bumpless transfer via SMRC approach

Notice that as the plant input with or without SMRC compensation is the same until $t = t_{switch}$, the time evolution of the controlled variable y coincides in Figures 8.2 and 8.4 until $t = t_{switch} + L_p$, where L_p is the plant delay.

8.4   MIMO bumpless transfer

We now move on to the development of a multivariable method for achieving bumpless transfers in MIMO systems [41]. To this end, SMRC ideas are also exploited. Since the existence of a reaching mode is not critical in the present application, the filter can now be placed outside the main control loop. We briefly describe the so-called collective sliding modes in the following text, as they are necessary for understanding the method operation. Note that these concepts are also useful for further developments on multivariable SMRC loops.

8.4.1   Some concepts on collective sliding modes

Consider the following dynamical system:

$$\dot{x} = Ax + \sum_{i=1}^{m} b_i w_i = Ax + Bw \tag{8.3}$$

where $x \in \mathbb{R}^n$ is the system state and $w \in \mathbb{R}^m$ is the generic control vector. Matrices A and B (and its column vectors b_i) are of consistent dimensions. A variable structure control law can be defined componentwise as

$$w_i = \begin{cases} w_i^+ & \text{if} \quad \sigma_i(x) > 0 \\ w_i^- & \text{if} \quad \sigma_i(x) < 0 \end{cases}, \quad i = 1, \ldots, m \tag{8.4}$$

according to the sign of the scalar switching functions $\sigma_i(x) = r_i - k_i^{\mathrm{T}} x$.

The sliding surface $\mathcal{S}$ is then defined as the intersection of the m individual sliding surfaces $\mathcal{S}_i$ defined by

$$\mathcal{S}_i = \{x \in \mathbb{R}^n : \sigma_i(x) = 0\} \tag{8.5}$$

Thus,

$$\mathcal{S} = \bigcap_{i=1}^{m} \mathcal{S}_i = \{x \in \mathbb{R}^n : \sigma(x) = r - K^{\mathrm{T}} x = 0\} \tag{8.6}$$

where $\sigma(x)$ is the vector of switching functions, i.e. $\sigma(x) = [\sigma_1(x) \ \sigma_2(x) \ \cdots \ \sigma_m(x)]^{\mathrm{T}}$. Besides, $r = [r_1 \ r_2 \ \cdots \ r_m]^{\mathrm{T}}$ is the SM reference vector. Finally, the columns of the matrix K are the feedback gain vectors k_i, i.e. $K = [k_1 \ k_2 \ \cdots \ k_m]$.

As studied in Chapter 2, a sliding motion exists locally on a particular individual sliding surface $\mathcal{S}_j$ if, as a result of the switching logic (8.4), the following reaching condition is satisfied:

$$\begin{cases} \dot{\sigma}_j < 0 & \text{if} \quad \sigma_j > 0 \\ \dot{\sigma}_j > 0 & \text{if} \quad \sigma_j < 0 \end{cases} \tag{8.7}$$

A necessary condition for (8.7) that is required to be satisfied is that the switching function σ_j should have relative degree one with respect to the discontinuous signal w_j.

Once the individual surface $\mathcal{S}_j$ is reached, the control action w_j switches at high frequency constraining the state trajectory to $\mathcal{S}_j$. If the individual sliding motion converges towards the intersection surface $\mathcal{S}$, then the combination of individual sliding motions results in a *collective sliding regime*. However, the intersection surface $\mathcal{S}$ can also be reached without arriving first at any individual surface $\mathcal{S}_j$.

Certainly, for the existence of collective sliding regimes on the intersection of the surfaces $\mathcal{S}_j$ ($\sigma_j(x) = 0, j = 1, \ldots, m$), it is not necessary that condition (8.7) holds true for each individual surface [136]. In this case, the existence conditions of MIMO sliding modes can be formulated in terms of the stability theory given by Lyapunov functions. This is established by Theorem 8.1.

Theorem 8.1: *A sufficient condition for the establishment of a collective SM on the intersection $\mathcal{S}$ of m surfaces $\mathcal{S}_j$ ($\sigma_j(x) = 0, j = 1, \ldots, m$), whose derivative function* *vector is given by*

$$\dot{\sigma} = d(x) - \eta D(x) sign(\sigma) \tag{8.8}$$

is that for $x \in \mathcal{S}^d \subset \mathcal{S}$, $\mathcal{S}^d$ the domain of interest in $\mathcal{S} = \{x : \sigma(x) = 0\}$

(c1) *$D(x)$ is positive definite, $D(x) + D^T(x) > 0$*   (8.9)

(c2) *$\eta > \sqrt{m} \, \| d(x) \|_2 / \lambda_{min}(x)$*   (8.10)

with $\eta \in \mathbb{R}^+$ and $\lambda_{min}(x)$ the minimum eigenvalue of $(D(x) + D^T(x))/2$.

Proof:

Let $V = |\sigma_1| + |\sigma_2| + \cdots + |\sigma_m| = sign(\sigma)^T \sigma = z^T \sigma$ be a Lyapunov function candidate. Assuming that SM is established on the intersection of k surfaces, we partition the vector σ into two subvectors $\sigma^T = [(\sigma^k)^T \quad (\sigma^{m-k})^T]$, so that $\sigma^k = 0 \in \mathbb{R}^k$, but $\sigma^{m-k} \neq 0 \in \mathbb{R}^{m-k}$.

According to the continuous equivalent control, vector $z^k = sign(\sigma^k)$ must be replaced by the function z_{eq}^k in the SM motion equation such that $\dot{\sigma}^k = 0 \in \mathbb{R}^k$. Hence, in SM the time derivative of V results

$$
\begin{aligned}
\dot{V} &= \frac{d}{dt}(z^T \sigma) = \frac{d}{dt}\left(\begin{bmatrix} z_{eq}^k \\ z^{m-k} \end{bmatrix}^T \begin{bmatrix} \sigma^k \\ \sigma^{m-k} \end{bmatrix} \right) \\
&= \begin{bmatrix} \dot{z}_{eq}^k \\ 0 \end{bmatrix}^T \begin{bmatrix} 0 \\ \sigma^{m-k} \end{bmatrix} + z^T \dot{\sigma} = z^T \dot{\sigma}
\end{aligned}
\tag{8.11}
$$

Replacing vector $\dot{\sigma}$ with its expression in (8.8) yields

$$\dot{V} = z^T d(x) - \eta z^T D(x) z \tag{8.12}$$

Since the components of vector z are either $sign(\sigma_i)$ or $z_{i_{eq}}$, with $|z_{i_{eq}}| \leq 1$, the first term in (8.12) can be upper bounded as

$$z^T d(x) \leq \| z \|_2 \| d(x) \|_2 \leq \sqrt{m} \, \| d(x) \|_2 \tag{8.13}$$

Besides, the second term in (8.12) is negative if matrix $D(x)$ is positive definite, in which case

$$-\eta z^T D(x) z = -\eta z^T \frac{D(x) + D^T(x)}{2} z \leq -\eta \lambda_{min}(x) \, \| z \|_2^2 \tag{8.14}$$

where $\| z \|_2 \geq 1$ because we assumed $\sigma^{m-k} \neq 0 \in \mathbb{R}^{m-k}$ initially.

Then, from (8.13) and (8.14), the time derivative of the Lyapunov function V can be upper bounded by

$$\dot{V} \leq \sqrt{m} \, \|d(x)\|_2 - \eta \lambda_{min}(x) \tag{8.15}$$

From (8.15), provided η fulfills (8.10), the Lyapunov function decays at a finite rate. Thus, the origin $\sigma = 0$ is an asymptotically stable equilibrium point with finite time convergence. This means that both V and σ vanish after a finite time interval and a collective SM is established on the intersection surface $\mathcal{S}$.

8.4.2   A MIMO bumpless algorithm

Figure 8.5 presents a schematic diagram of a control system with the manual–automatic *bumpless* strategy based on SMRC ideas. A slightly changed scheme will be suggested in Section 8.5 to address commutation between different controllers, for which most of the following analysis is also valid.

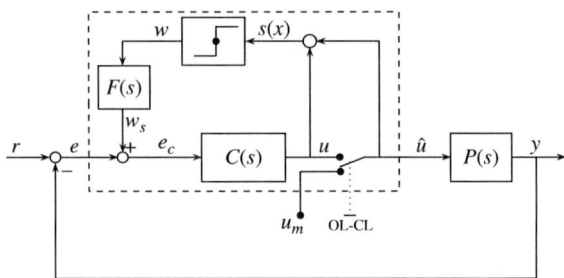

Figure 8.5   Dynamic SM bumpless *strategy for multivariable systems*

$P(s) \in \mathbb{R}^{m \times m}(s)$ represents the plant to be controlled, whereas $C(s) \in \mathbb{R}^{m \times m}(s)$ is a multivariable controller (either centralised or decentralised) designed for stable closed-loop operation around the operating point P_0. It is assumed that this controller is biproper, thus it has a minimal realisation like

$$C(s): \begin{cases} \dot{x}_c = A_c x_c + B_c e_c \\ u = C_c x_c + D_c e_c \end{cases} \tag{8.16}$$

where D_c is a non-singular matrix. Observe that this assumption – which significantly simplifies the method explanation – does not impose a severe restriction, since in multivariable design biproper controllers are aimed to avoid unnecessary delays in closed-loop transfer matrices. In any way, the method can be extended to deal with strictly proper controllers by simply including additional controller states in the sliding functions, as was done for strictly proper SMRC algorithms along the book.

The loop around $C(s)$, inside the dotted box in Figure 8.5, is the proposed compensation, which can be seen as a dynamic MIMO-SM block [115]. Therein, the filter $F(s)$ is aimed to smooth the signal w_s added at the controller input and to guarantee the transversality condition for the establishment of sliding regimes. This filter can be represented in state space as

$$F(s) : \begin{cases} \dot{x}_f = A_f x_f + B_f w \\ w_s = C_f x_f \end{cases} \tag{8.17}$$

where $A_f = -C_f = \lambda_f I_m$ and $B_f = I_m$.

The discontinuous signal $w \in \mathbb{R}^m$ is given by

$$w(x) = M(x) sign(\sigma(x)) \tag{8.18}$$

where $x = [x_c^{\mathrm{T}} \ x_f^{\mathrm{T}}]^{\mathrm{T}}$, $M(x)$ is a diagonal gain matrix and

$$\sigma(x) = [\sigma_1(x) \quad \sigma_2(x) \quad \cdots \quad \sigma_m(x)]^{\mathrm{T}} = \hat{u} - u(x_c, e_c) \tag{8.19}$$

is the switching function, with $u \in \mathbb{R}^m$ and $\hat{u} \in \mathbb{R}^m$ vectors that contain the controller outputs and plant inputs, respectively.

The objective of the variable structure loop added to the original controller $C(s)$ is the establishment of a sliding regime on the surface $S = \{x : \sigma(x) = 0\}$, in such a way that the controller outputs u are forced to coincide with the plant inputs $\hat{u}$, thus avoiding jumps at the plant input when the main control loop is closed. Naturally, in order to guarantee the stability of the closed-loop system, commutation must be performed within the domain of attraction [108] of the operating point P_0, which may be characterised as

$$D_0 = \{x_{cl_0} : \lim_{t \to \infty} \phi(t, t_0, x_{cl}) = P_0\} \tag{8.20}$$

where x_{cl} is the closed-loop state vector and $\phi(t, t_0, x_{cl})$ is the state trajectory corresponding to the initial condition $x_{cl}(t_0) = x_{cl_0}$ evaluated at time t.

8.4.2.1   Reaching condition

In the proposed *bumpless* scheme, replacing the expressions of (8.16) and (8.17) in the derivative of the switching function $\sigma = \hat{u} - u$ and comparing with (8.8) yields

$$\begin{aligned} d(x) &= \dot{\hat{u}} - C_c A_c x_c - C_c B_c e_c - D_c \dot{e} - D_c C_f A_f x_f \\ &= \dot{\hat{u}} - C_c A_c x_c - C_c B_c e_c - D_c \dot{e} - \lambda_f D_c w_s \end{aligned} \tag{8.21}$$

$$\begin{aligned} D(x) &= \frac{D_c C_f B_f M}{\eta} \\ &= \frac{-\lambda_f D_c M}{\eta} \end{aligned} \tag{8.22}$$

Therefore, SM establishment will be assured provided $d(x)$ and $D(x)$ satisfy conditions **c1** and **c2** in (8.9) and (8.10), respectively. Clearly, the matrix $D(x)$ depends on the control law (8.18), and consequently the condition **c1** can be satisfied by choosing the signs of the diagonal entries in M so that every leading principal minor of M is positive [18]. Moreover, the value $\lambda_{min}(x)$ can be incremented by enlarging the gains in M. This allows, once verified condition **c1**, adjusting these diagonal entries in order to satisfy the reaching condition **c2**. In this manner, the SM establishment will be assured, and consequently the error between the controller output vector u and the plant input $\hat{u}$ will reduce to zero, thereby eliminating potential jumps in the signal at the plant input when commutation is performed. It is important to remember that, from a theoretical point of view, for bounded initial conditions the time in which the system reaches the surface S (*reaching time*) can be made arbitrarily short [136].

Remark 8.1: *As mentioned, a positive definite $D(x)$ can be obtained by a proper choice of the entry signs in M. However, for high-dimensional multivariable systems, selecting each entry sign of the matrix M after inspection may become tedious. A simple way of assuring condition **c1** for this kind of systems is to take*

$$M = \eta D_c^{-1}, \quad \eta \in \mathbb{R}^+ \tag{8.23}$$

so that

$$D(x) = -\lambda_f I_m \tag{8.24}$$

which is always positive definite.

8.4.2.2 Hidden SM dynamics

An open-loop representation of the SM compensation can be derived from the state-space representations of the controller $C(s)$ and the filter $F(s)$, (8.16) and (8.17), respectively, resulting in

$$\begin{bmatrix} \dot{x}_c \\ \dot{e}_c \end{bmatrix} = \begin{bmatrix} A_c & B_c \\ 0 & \lambda_f \end{bmatrix} \begin{bmatrix} x_c \\ e_c \end{bmatrix} + \begin{bmatrix} 0 \\ \dot{r} - \dot{y} - \lambda_f(r - y) \end{bmatrix} + \begin{bmatrix} 0 \\ -\lambda_f w \end{bmatrix} \tag{8.25}$$

$$u = C_c x_c + D_c e_c \tag{8.26}$$

Since $C(s)$ is biproper (D_c is non-singular), during SM

$$e_c = D_c^{-1}(\hat{u} - C_c x_c) \tag{8.27}$$

which results from (8.26) after making (8.19) equal to zero. Thus, the last row of (8.25) becomes redundant. Replacing (8.27) in the first row of (8.25) results in the already familiar SM reduced dynamics

$$\dot{x}_c = Q_c x_c + B_c D_c^{-1} \hat{u}$$
$$Q_c = A_c - B_c D_c^{-1} C_c \tag{8.28}$$

where the eigenvalues of Q_c are the zeros of the controller $C(s)$. Therefore, the SM dynamics given by (8.28) will be stable as long as the controller is minimum phase.

It is important to remark that this SM dynamics does only depend on the controller parameters, and it is not seen from the controller output because $u = \hat{u}$ during the sliding regime. Thus, all the dynamics associated with SM is hidden dynamics.

Another interesting feature of the proposed compensation is that the second term of (8.25) satisfies the matching condition. Consequently, because of SM robustness properties, the variable structure loop is invariant to the references r and the output disturbances that might appear in y. This is verified by (8.28).

8.5   Application to the *quadruple tank* process

The bumpless algorithm is evaluated in the benchmark *quadruple tank* process, as an example of non-linear MIMO processes. The non-linear model of this plant was already presented in Chapter 6 (see (6.44)).

The values of the system parameters for the operating point $P-$, at which the system shows minimum-phase characteristics, are given in Table 6.1. For this operating point, a controller that achieves full dynamic decoupling was designed following the procedures described in Chapter 4. The resulting centralised controller consists of the following individual transfer functions:

$$c_{11}(s) = \frac{2.385(s+0.043)(s+0.033)(s+0.016)}{s(s+0.059)(s+0.017)} \tag{8.29}$$

$$c_{12}(s) = \frac{-0.08(s+0.033)(s+0.011)}{s(s+0.059)(s+0.017)} \tag{8.30}$$

$$c_{21}(s) = \frac{-0.039(s+0.043)(s+0.016)}{s(s+0.059)(s+0.017)} \tag{8.31}$$

$$c_{22}(s) = \frac{3.214(s+0.043)(s+0.033)(s+0.011)}{s(s+0.059)(s+0.017)} \tag{8.32}$$

and it has all its multivariable zeros in the left-half plane.

8.5.1   *Manual–automatic switching*

The non-linear system was taken manually around the desired operating point, and once sufficiently close to $P-$ the controller given by (8.29)–(8.32) was connected. The time instant for commutation to automatic mode (t_{switch}) was set as the one in which both water levels surpass the 90% of the value corresponding to $P-$. Figure 8.6 plots the results obtained by simulations carried out without the bumpless compensation. The jumps at the plant inputs caused by the inconsistency between u and $\hat{u}$ were limited to 12 V with an amplitude limiter, but even then they produced unacceptable transients in the water levels whose control is aimed.

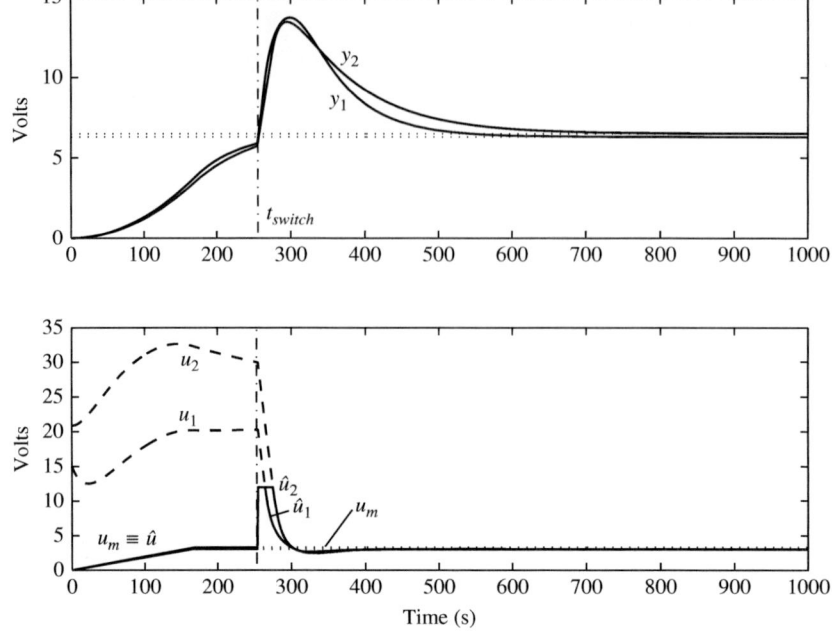

*Figure 8.6   Jumps at the plant inputs and undesired transients in the controlled
variables because of uncompensated manual–automatic
commutation*

Certainly, if this commutation had been performed on the real system, it would
have overflowed the lower tanks.

To improve the system response, the SM algorithm was added. Taking
$\lambda_f = -0.05$ in $F(s)$ and $M = 20I_2$, from (8.29) to (8.32) we have

$$D(x) = D_c = \begin{bmatrix} 2.385 & 0 \\ 0 & 3.214 \end{bmatrix} \tag{8.33}$$

which is clearly positive definite.

The responses obtained with the dynamic SM compensation of the MIMO
controller are shown in Figure 8.7. It reveals how the controlled variables reach the
desired operating point without overshoots after the mode switching. The figure
also verifies that the controller output u is always close to u_m during manual
operation (until t_{switch}). It is worth highlighting that although the *chattering* in u
previous to commutation can be made negligible as the SM is confined to the low-
power side of the system, this phenomenon would not affect any component or
variable of the system. Actually, the chattering in u may only occur when the
controller is not connected to the plant.

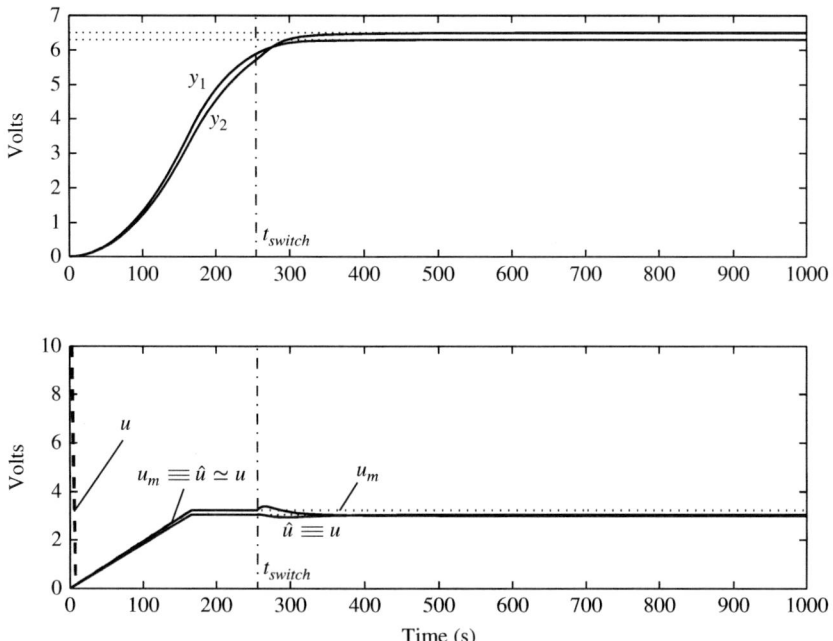

Figure 8.7   Multivariable bumpless *transfer via MIMO-SM method*

8.5.2   *Automatic–automatic commutation*

The SMRC methodology is now applied to avoid bumpy transfers when switching between different controllers is carried out. With this objective, the scheme shown in Figure 8.5 is slightly modified as shown in Figure 8.8 for the case of the quadruple tank.

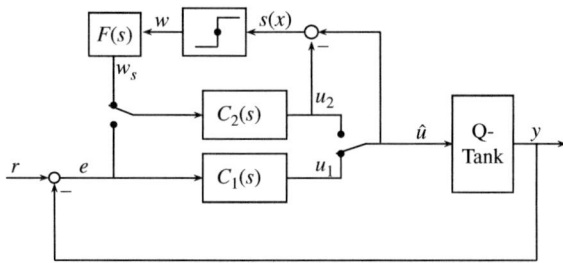

Figure 8.8   MIMO bumpless topology for controller switching

Observe that the analysis of the previous section is also valid for this configuration, except for the fact that (8.25) will also depend on controller $C_1(s)$. However, it is worth mentioning that, like it occurs in general with bumpless algorithms when non-arbitrary switching is considered (see References 131, 155 and references therein), the present proposal does not guarantee the stability of the switched closed-loop system. This depends on the conditions under which controllers are switched, and it might be considered for instance in a supervisory level. Even so, the proposed method reduces the risk of instability by making the off-line controller states consistent with the actual plant inputs.

Controller $C_1(s)$ in Figure 8.8 corresponds to the one described by (8.29)–(8.32) that was designed to decouple the system around $P-$, whereas controller $C_2(s)$ is a decoupler controller obtained from the model linearisation at another operating point P^*-, with $(v_1^0; v_2^0) = (3.9; 3.9)$ V and $(h_1^0; h_2^0; h_3^0; h_4^0) = (20.7; 21.6; 2.8; 2.4)$ cm. To preserve a good degree of decoupling of the non-linear system, a commutation to $C_2(s)$ is scheduled for the case in which both controlled water levels h_1 and h_2 are greater than 17 cm (recall that, according to the sensor sensitivity, this corresponds to 8.5 V). $C_1(s)$ and $C_2(s)$ has the same relative degree structure and direct feedforward matrix D_c, but different poles and zeros locations. Therefore, the SM compensation can be implemented with the same filter $F(s)$ and matrix M that were used for the manual–automatic switching.

Once at $P-$ (it can be viewed as being at $t = 1000$ s in the top plot given in Figure 8.7), two positive step references are applied to take the system to P^*-, and as a consequence the condition for controller switching is reached at $t = 1306$ s. The responses to these step references and the corresponding controller commutation are depicted in Figure 8.9 for both the uncompensated system (dashed lines)

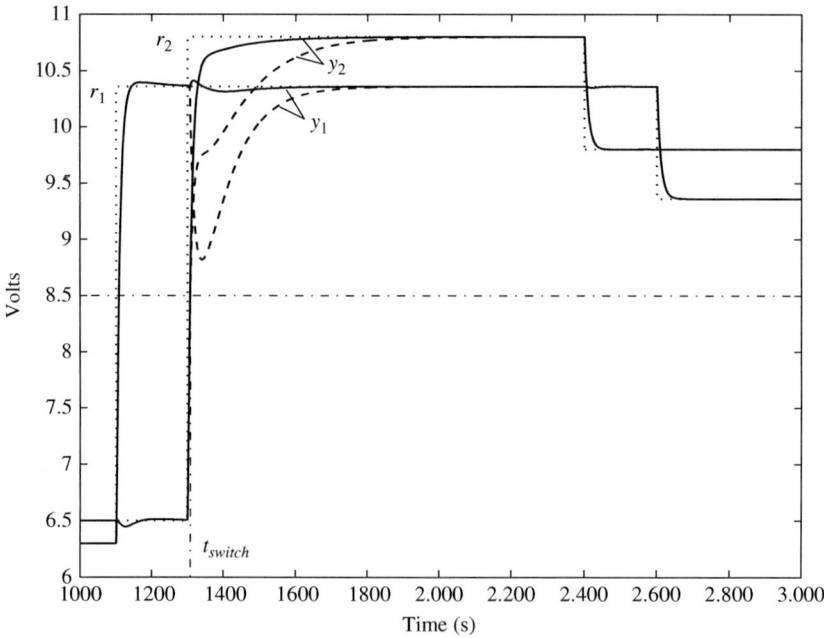

Figure 8.9   Bumpy (dashed) and bumpless (solid) transfer between decoupling controllers for the quadruple tank at different operating points

and the compensation of Figure 8.8 (solid lines). The MIMO-SM bumpless compensation considerably reduces the transient caused by the controller switching, and it helps to preserve the decoupling of the system. The closed-loop response against the two last (negative) step references verifies that $C_2(s)$ achieves the dynamic decoupling of the quadruple tank at P^*-.

References

1. H. Amann. *Ordinary Differential Equations*. 1st edition. Walter de Gruyter, Berlin, 1990.
2. J. Angeles. *Fundamentals of Robotic Mechanical Systems: Theory, Methods, and Algorithms*. 3rd edition. Springer-Verlag, New York, NJ, 2007.
3. K. Åström, K. Johansson, Q. Wang. 'Design of decoupled PID controllers for two-by-two systems'. *IEE Proceedings on Control Theory and Applications*, **149**:74–81, 2002.
4. K. J. Åström, T. Hägglund. *Advanced PID control*. ISA (Instrumentation, System and Automation Society), 2006.
5. K. J. Åström, L. Rundqwist. 'Integrator windup and how to avoid it'. *Proceedings of the American Control Conference*, pp. 1693–1698, Pittsburgh, 1989.
6. G. Bartolini, A. Pisano, E. Punta, E. Usai. 'A survey of applications of second-order sliding mode control to mechanical systems'. *International Journal of Control*, **76**:875–892, 2003.
7. G. Bastin, D. Dochain. *On-line Estimation and Adaptive Control of Bioreactors*. Elsevier, Amsterdam, The Netherlands, 1990.
8. H. De Battista, J. Picó, E. Picó-Marco. 'Globally stabilizing control of fedbatch processes with Haldane kinetics using growth rate estimation feedback'. *Journal of Process Control*, **16**:865–875, 2006.
9. H. De Battista, J. Picó, E. Picó-Marco, V. Mazzone. 'Adaptive sliding mode control of fed-batch processes using specific growth rate estimation feedback'. *Proceedings of 10th IFAC Computer Applications in Biotechnology*, Cancun, Mexico, 2007.
10. F. Bianchi, H. De Battista, R. Mantz. *Wind Turbine Control Systems*. Springer, London, 2006.
11. T. S. Brinsmead, G. C. Goodwin. 'Cheap decoupled control'. *Automatica*, **37**(9):1465–1471, 2001.
12. E. Bristol. 'On a new measure of interaction for multivariable process control'. *IEEE Transactions on Automatic Control*, **11**:133–134, 1966.
13. H. Bühler. *Regelkreise mit Begrenzungen*, Vol. 828. VDI-Verlag, Düsseldorf, 2000.
14. T. Burton, D. Sharpe, N. Jenkins, E. Bossanyi. *Wind Energy Handbook*. John Wiley and Sons, Chichester, 2001.
15. P. Camocardi, P. Battaiotto, R. Mantz. 'Brushless doubly fed induction machine in wind generation for water pumping'. *Proceedings of the*

International Conference on Electrical Machines 2008 (ICEM08), ISBN: 978-1-4244-1736-0, 2008.

16. P. Campo, M. Morari. 'Robust control of processes subject to saturation nonlinearities'. *Computers and Chemical Engineering*, **14**:343–358, 1990.

17. P. Campo, M. Morari. 'Achievable closed-loop properties of systems under decentralized control: conditions involving the steady-state gain'. *IEEE Transactions on Automatic Control*, **39**(5):932–943, 1994.

18. C. Chen. *Linear System Theory and Design*. 3rd edition. Oxford University Press, New York, 1999.

19. A. Conley, M. Salgado. 'Gramian based interaction measure'. *Proceedings of the Conference on Decision and Control*, pp. 5020–5022, Sydney, Australia, December 2000.

20. H. De Battista, R. Mantz, F. Garelli. 'Power conditioning for a wind-hydrogen energy system'. *Journal of Power Sources*, **155**:478–486, 2006.

21. R. M. Dell, D. A. Rand. 'Energy storage – a key technology for global energy sustainability'. *Journal of Power Sources*, **100**(1):2–17, 2001.

22. C. A. Desoer, A. N. Gündes. 'Decoupling linear multiinput multioutput plants by dynamic output feedback: an algebraic theory'. *IEEE Transactions on Automatic Control*, **31**:744–750, 1986.

23. L. Dewasme, F. Renard, A. Vande Wouwer. 'Experimental investigations of a robust control strategy applied to cultures of *S. cerevisiae*'. *Proceedings of the European Control Conference*, Kos, Greece, 2007.

24. A. G. Dutton, J. A. M. Bleijs, H. Dienhart, M. Falchetta, W. Hug, D. Prischich, *et al.* 'Experience in the design, sizing, economics, and implementation of autonomous wind-powered hydrogen production systems'. *International Journal of Hydrogen Energy*, **25**(8):705–722, 2000.

25. C. Edwards, I. Postlethwaite. 'Anti-windup and bumpless transfer schemes'. *Automatica*, **34**(2):199–210, 1998.

26. C. Edwards, S. K. Spurgeon. *Sliding Mode Control: Theory and Applications*. 1st edition. Taylor & Francis, UK, 1998.

27. C. Elam, B. Kroposki, G. Bianchi, K. Harrison. 'Renewable electrolysis integrated system development and testing'. Progress Report FY2004, DOE Hydrogen Program, NREL, Colorado, 2004.

28. V. Eldem. 'The solution of diagonal decoupling problem by dynamic output feedback and constant precompensator: the general case'. *IEEE Transactions on Automatic Control*, **39**(3):503–511, 1994.

29. S. Emel'yanov. *Automatic Control Systems of Variable Structure*. Russian edition. Nauka, Moscow, 1967.

30. F. Esfandiari, H. K. Khalil. 'Stability analysis of a continuous implementation of variable structure control'. *IEEE Transactions on Automatic Control*, **36**(5):616–620, 1991.

31. L. Fingersh. 'Optimized hydrogen and electricity generation from wind'. Technical Report TP-500-34364, NREL, Colorado, June 2003.

32. L. Fridman, A. Levant. *Sliding Mode Control in Engineering*. Marcel Dekker, Inc., 2002.

33. C. E. Garcia, M. Morari. 'Internal model control-1. A unifying review and some new results'. *Industrial Engineering Chemical Process Design and Development*, **21**:308–323, 1982.

34. C. E. Garcia, M. Morari. 'Internal model control-2. Design procedure for multivariable systems'. *Industrial Engineering Chemical Process Design and Development*, **24**:472–484, 1985.

35. C. E. Garcia, M. Morari. 'Internal model control-3. Multivariable control law computation and tuning guidelines'. *Industrial Engineering Chemical Process Design and Development*, **24**:484–494, 1985.

36. F. Garelli, P. Camocardi, R. Mantz. 'Variable structure strategy to avoid amplitude and rate saturation in pitch control of a wind turbine'. *International Journal of Hydrogen Energy*, **35**:5869–5875, 2010.

37. F. Garelli, L. Gracia, A. Sala, P. Albertos. 'Sliding mode speed auto-regulation technique for robotic tracking'. *Robotic and Autonomous Systems*, **59**:519–529, 2011.

38. F. Garelli, R. Mantz, H. De Battista. 'Limiting interactions in decentralized control of MIMO systems'. *Journal of Process Control*, **16**(5):473–483, 2006.

39. F. Garelli, R. Mantz, H. De Battista. 'Partial decoupling of non-minimum phase processes with bounds on the remaining coupling'. *Chemical Engineering Science*, **61**:7706–7716, 2006.

40. F. Garelli, R. Mantz, H. De Battista. 'Sliding mode compensation to preserve decoupling of stable systems'. *Chemical Engineering Science*, **62**:4705–4716, 2007.

41. F. Garelli, R. Mantz, H. De Battista. 'Collective sliding mode technique for multivariable bumpless transfer'. *Industrial and Engineering Chemistry Research*, **47**:2721–2727, 2008.

42. F. Garelli, R. Mantz, H. De Battista. 'Multi-loop two-degrees-of-freedom PI controller with adaptive set-point weighting'. *Journal of Systems and Control Engineering*, **224**(8):1033–1039, 2010.

43. F. Garelli, M. Ramírez, A. Domínguez, A. Angulo. 'Simulation of an algorithm for chute level control under discontinuous cane supply'. *Revista Iberoamericana de Automática Industrial*, **6**(3):54–60, 2009.

44. E. Gilbert, K. Tan. 'Linear systems with state and control constraints: the theory and application of maximal output admissible sets'. *IEEE Transactions on Automatic Control*, **36**(9):1008–1020, 1991.

45. O. Gjosaeter, A. F. Bjarne. 'On the use of diagonal control versus decoupling for ill-conditioned processes'. *Automatica*, **33**:427–432, 1997.

46. A. Glattfelder, W. Schaufelberger. *Control Systems with Input and Output Constraints*. Springer, London, 2003.

47. G. Gómez, G. Goodwin. 'Using coprime factorizations in partial decoupling of linear multivariable systems'. *Proceedings of the American Control Conference*, pp. 2053–2057, 1997.

48. G. Goodwin, A. Feuer, G. Gómez. 'A state-space technique for evaluation of diagonalizing compensators'. *Systems and Control Letters*, **32**(3):173–177, 1997.

49. G. Goodwin, S. Graebe, M. Salgado. *Control System Design*. 1st edition. Prentice Hall, New Jersey, 2001.
50. G. Goodwin, M. Salgado, E. Silva. 'Time-domain performance limitations arising from decentralized architectures and their relationship to the RGA'. *International Journal of Control*, **78**(13):1045–1062, 2005.
51. G. Goodwin, M. Seron, J. de Doná. *Constrained Control and Estimation. An Optimisation Approach*. Springer, London, 2005.
52. P. Grosdidier, M. Morari, B. Holt. 'Closed-loop properties from steady-state gain information'. *Industrial and Engineering Chemistry Fundamentals*, **24**(2):221–235, 1985.
53. C. C. Hang, K. J. Åström, Q. G. Wang. 'Relay feedback auto-tuning of process controllers – a tutorial review'. *Journal of Process Control*, **12**:143–162, 2002.
54. R. Hanus, M. Kinnært. 'Control of constrained multivariable systems using the conditioning technique'. *Proceedings of the American Control Conference*, pp. 1712–1718, Pittsburgh, 1989.
55. R. Hanus, M. Kinnaert, J. Henrotte. 'Conditioning technique, a general anti-windup and bumpless transfer method'. *Automatica*, **23**(6):729–739, 1987.
56. E. Hau. *Wind Turbines: Fundamentals, Technologies, Applications, Economics*. Springer, London, 2006.
57. M. Hautus, M. Heymann. 'Linear feedback decoupling-transfer function analysis'. *IEEE Transactions on Automatic Control*, **28**(8):823–832, 1983.
58. S. Heier. *Grid Integration of Wind Energy Conversion Systems*. John Wiley & Sons Ltd., Chichester, England, 1998.
59. B. Henes, B. B. Sonnleitner. 'Controlled fed-batch by tracking the maximal culture capacity'. *Journal of Biotechnology*, **132**:118–126, 2007.
60. P. Hippe. *Windup in Control. Its effects and their prevention*. Advances in Industrial Control. Springer, London, 2006.
61. M. Hovd, S. Skogestad. 'Simple frequency-dependent tools for control system analysis, structure selection and design'. *Automatica*, **28**(5):989–996, 1992.
62. T. Hu, Z. Li. *Control Systems with Actuator Saturation: Analysis and Design*. Birkhäuser, Boston, 2001.
63. M. Huba, S. Skogestad, M. Fikar, M. Hovd, T. Johansen, B. Rohal'-Ilkiv (eds.). *Selected Topics on Constrained and Nonlinear Control*. STU Bratislava – NTNU Trondheim, 2011.
64. J. Hung, W. Gao, J. C. Hung. 'Variable structure control: a survey'. *IEEE Transactions on Industrial Electronics*, **40**(1):2–22, 1993.
65. A. Isidori. *Nonlinear Control Systems*. 3rd edition. Springer, London, 1995.
66. K. Johansson. 'The quadruple-tank process: a multivariable laboratory process with an adjustable zero'. *IEEE Transactions on Control Systems Technology*, **8**(3):456–465, 2000.
67. K. Johansson. 'Interaction bounds in multivariable control systems'. *Automatica*, **38**(6):1045–1051, 2002.

68. S. Kanev, T. Engelen. 'Exploring the limits in individual pitch control'. *Proceedings of European Wind Energy Conference 2009*, pp. 1–12, 2009.

69. V. Kapila, K. Grigoriadis. *Actuator Saturation Control*. CRC Press, New York, 2002.

70. A. Khaki-Sedigh, B. Moaveni. *Control Configuration Selection for Multivariable Plants*. Springer-Verlag, Berlin, 2009.

71. K. Kivijarju, K. Salonen, U. Moilanen, E. Meskanen, M. Leisola, T. Eerikainen. 'On-line biomass measurements in bioreactor cultivations: comparison study of two on-line probes'. *Journal of Industrial Microbiology and Biotechnology*, **34**(8):561–566, 2007.

72. M. Kothare, P. Campo, M. Morari, K. Nett. 'A unified framework for the study of anti-windup designs'. *Automatica*, **30**(12):1869–1883, 1994.

73. F. N. Koumboulis. 'Input-output triangular decoupling and data sensitivity'. *Automatica*, **32**:569–573, 1996.

74. F. N. Koumboulis, M. G. Skarpetis. 'Robust triangular decoupling with application to 4WS cars'. *IEEE Transactions on Automatic Control*, **45**(2):344–352, 2000.

75. J. Lee, T. Edgar. 'Interaction measure for decentralized control of multivariable processes'. *Proceedings of the American Control Conference*, pp. 454–458, Anchorage, AK, May 2002.

76. A. Levant. 'Higher order sliding: differentiation and black-box control'. *Proceedings of IEEE Conference on Decision and Control*, pp. 1703–1708, 2000.

77. A. Levant. 'Universal SISO sliding-mode controllers with finite time convergence'. *IEEE Transactions on Automatic Control*, **46**:1447–1451, 2001.

78. A. Levant. 'Higher-order sliding modes, differentiation and output-feedback control'. *International Journal of Control*, **76**(9/10):924–941, 2003.

79. D. Liberzon. *Switching in Systems and Control*. Birkhäuser, Boston, 2003.

80. C.-A. Lin. 'Necessary and sufficient conditions for existence of decoupling controllers'. *IEEE Transactions on Automatic Control*, **42**(8):1157–1161, 1997.

81. C.-A. Lin, T.-F. Hsieh. 'Decoupling controller design for linear multivariable plants'. *IEEE Transactions on Automatic Control*, **36**(4):485–489, 1991.

82. A. Linnemann, R. Maier. 'Decoupling by precompensation while maintaining stabilizability'. *IEEE Transactions on Automatic Control*, **38**(4):629–632, 1993.

83. A. G. J. MacFarlane, N. Karcanias. 'Poles and zeros of linear multivariable systems: a survey of algebraic, geometric and complex variable theory'. *International Journal of Control*, **24**:33–74, 1976.

84. R. Mantz, H. De Battista, F. Bianchi. 'VSS global performance improvement based on AW concepts'. *Automatica*, **41**(6):1099–1103, 2005.

85. J. Mareczek, M. Buss, M. Spong. 'Invariance control for a class of cascade nonlinear systems'. *IEEE Transactions on Automatic Control*, **47**(4): 636–640, 2002.

86. T. J. McAvoy. *Interaction Analysis: Principles and Applications*. ISA, Research Triangle Park, North Carolina, 1983.

87.   N. Mohan, T. M. Undeland, W. P. Robbins. *Power Electronics: Converters, Applications and Design.* 3rd edition. John Wiley & Sons, New York, 2003.

88.   M. Morari, E. Zafiriou. *Robust Process Control.* Prentice Hall, New Jersey, 1989.

89.   A. S. Morse, W. M. Wonham. 'Triangular decoupling of linear multivariable systems'. *IEEE Transactions on Automatic Control*, **15**:447–449, 1970.

90.   E. Mulder, M. Kothare, M. Morari. 'Multivariable anti-windup controller synthesis using LMI'. *Automatica*, **37**:1407–1416, 2001.

91.   J. L. Navarro, D. Barreras, J. Picó, E. Picó-Marco, J. Romero. 'A new sensor for absorbance measurement'. *Proceedings of IFAC 9th International Symposium on Control Applications on Biotechnology*, Nancy, France, 2004.

92.   C. Nett, V. Manousiouthakis. 'Euclidean condition and block relative gain: connections, conjectures and clarifications'. *IEEE Transactions on Automatic Control*, **5**:405–407, 1987.

93.   T. Ozkocak, M. Fu, G. Goodwin. 'A nonlinear modelling approach to the sugar cane crushing process'. *Proceedings of the 37th IEEE Conference on Decision and Control*, pp. 3144–3149, Tampa, Florida, USA, 1998.

94.   P. N. Paraskevopoulos, N. Koumboulis, N. D. Kouvakas, C. Balafas. 'I/O decoupling via dynamic controllers – a state space approach'. *Proceedings of the 13th Mediterranean Conference on Control and Automation*, Limassol, Cyprus, June 2005.

95.   Y. Peng, D. Vrancic, R. Hanus, S. Weller. 'Anti-windup design for multivariable controllers'. *Automatica*, **34**:1559–1565, 1998.

96.   Y. Peng, D. Vrančić, R. Hanus. 'Anti-windup, bumpless, and CT techniques for PID controllers'. *IEEE Control Systems Magazine*, **16**(4):48–56, 1996.

97.   J. Picó, F. Garelli, H. De Battista, R. Mantz. 'Geometric invariance and reference conditioning ideas for control of overflow metabolism'. *Journal of Process Control*, **19**:1617–1626, 2009.

98.   E. Picó-Marco, J. Picó, H. De Battista. 'Sliding mode scheme for adaptive specific growth rate control in biotechnological fed-batch processes'. *International Journal of Control*, **78**(2):128–141, 2005.

99.   J. Pires. *Industrial Robots Programming: Building Applications for the Factories of the Future.* Springer-Verlag, Berlin, 2007.

100. P. Puleston, R. Mantz. 'An anti-windup proportional integral structure for controlling time-delayed multiinput-multioutput processes'. *Industrial Engineering and Chemistry Research*, **34**:2993–3000, 1995.

101. P. Ratledge, B. Kristiansen. *Basic Biotechnology.* 2nd edition. Cambridge University Press, 2001.

102. J. Rawlings. 'Tutorial overview of model predictive control'. *IEEE Control Systems Magazine*, **20**(3):38–52, 2000.

103. F. Renard, A. Vande Wouwer, S. Valentinotti, D. Dumur. 'A practical robust control scheme for yeast fed-batch cultures – an experimental validation'. *Journal of Process Control*, **16**:855–864, 2006.

104. P. Roberts. *A study of brushless doubly-fed (induction) machines.* PhD thesis, Emmanuel College, University of Cambridge, 2004.

105. M. Rogozinsky, P. Paplinsky, M. Gibbard. 'An algorithm for the calculation of a nilpotent interactor matrix for linear multivariable systems'. *IEEE Transactions on Automatic Control*, **32**:234–237, 1987.

106. H. Rosenbrock. *State-Space and Multivariable Theory*. Nelson, London, 1970.

107. A. Saberi, A. Stoorvogel, P. Sannuti. *Control of Linear Systems with Regulation and Input Constraints*. Springer, London, 2000.

108. S. Sastry. *Nonlinear Systems: Analysis, Stability and Control*. Springer-Verlag, New York, 1999.

109. T. Schucan. *Case studies of integrated hydrogen energy systems*. Final report, International Energy Agency Hydrogen Implementing Agreement, Paul Scherrer Institute, Switzerland, 2000.

110. M. Seron, J. Braslavsky, G. Goodwin. *Fundamental Limitations in Filtering and Control*. Springer, London, 1997.

111. B. D. Shakya, L. Aye, P. Musgrave. 'Technical feasibility and financial analysis of hybrid wind-photovoltaic system with hydrogen production for Cooma'. *International Journal of Hydrogen Energy*, **30**(1):9–20, 2005.

112. Y. Shimon (Editor). *Handbook of Industrial Robotics*. Wiley, New York, NJ, 1999.

113. E. I. Silva, M. E. Salgado. 'Performance bounds for feedback control of non-minimum phase MIMO systems with arbitrary delay structure'. *IEE Proceedings – Control Theory and Applications*, **152**:211–219, 2005.

114. H. Sira-Ramírez. 'Differential geometric methods in variable structure systems'. *International Journal of Control*, **48**(4):1359–1390, 1988.

115. H. Sira-Ramírez. 'On the dynamical sliding mode control of nonlinear systems'. *International Journal of Control*, **57**(5):1039–1061, 1993.

116. S. Skogestad, M. Morari. 'Robust performance of decentralized control systems by independent designs'. *Automatica*, **25**(1):119–125, 1989.

117. S. Skogestad, M. Morari. 'Variable selection for decentralized control'. *Model, Identification and Control*, **13**(2):113–125, 1992.

118. S. Skogestad, I. Postlethwaite. *Multivariable Feedback Control: Analysis and Design*, 2nd edition. Wiley, Chichester, 2005.

119. J.-J. Slotine, S. S. Sastry. 'Tracking control of non-linear systems using sliding surfaces'. *International Journal of Control*, **38**:465–492, 1983.

120. I. Smets, G. Bastin, J. Van Impe. 'Feedback stabilization of fed-batch bioreactors: non-monotonic growth kinetics'. *Biotechnology Progress*, **18**:1116–1125, 2002.

121. S. Sojoudi, J. Lavaei, A. Aghdam. *Structurally Constrained Controllers*. Springer, New York, 2010.

122. B. Sonnleitner, O. Käpeli. 'Growth of *Saccharomyces cerevisiae* is controlled by its limited respiratory capacity: formulation and verification of a hypothesis'. *Biotechnology and Bioengineering*, **28**:927–937, 1986.

123. A. Stoorvogel, A. Saberi (eds.). Special Issue. 'Control problems with constraints'. *International Journal on Robust and Nonlinear Control*, **9**(10):583–734, 1999.

124. J. Chen, R. Middleton (eds.). Special Issue. 'New developments and applications in performance limitation of feedback control'. *IEEE Transactions on Automatic Control*, **48**(8):1297–1393, 2003.

125. M. Turner, L. Zaccarian (eds.). Special Issue. 'Anti-windup'. *International Journal of Systems Science*, **37**(2):65–139, 2006.

126. H. Sussmann, E. Sontag, Y. Yang. 'A general result on the stabilization of linear systems using bounded controls'. *IEEE Transactions on Automatic Control*, **39**:2411–2425, 1994.

127. S. Tarbouriech, G. Garcia, A. Glattfelder (eds.). 'Advanced Strategies in Control Systems with Input and Output Constraints'. Vol. 346. *Lecture Notes in Control and Information Sciences*. Springer, Berlin, 2007.

128. S. Tarbouriech, M. Turner. 'Anti-windup design: an overview of some recent advances and open problems'. *IET Control Theory and Applications*, **3**(1):1–19, 2009.

129. A. Teel, N. Kapoor. 'The L2 anti-windup problem: its definition and solution'. *Proceedings of 4th European Control Conference*, Brussels, Belgium, 1997.

130. M. Titica, D. Dochain, M. Guay. 'Adaptive extremum seeking control of fed-batch bioreactors'. *European Journal of Control*, **9**:618–631, 2003.

131. M. Turner, D. Walker. 'Linear quadratic bumpless transfer'. *Automatica*, **36**:1089–1101, 2000.

132. Ø. Ulleberg. 'Modeling of advanced alkaline electrolyzers: a system simulation approach'. *International Journal of Hydrogen Energy*, **28**(1):21–33, January 2003.

133. V. Utkin. *Sliding Regimes and their Application in Variable-Structure Systems*. Russian edition. Nauka, Moscow, 1974.

134. V. Utkin, H. Lee. 'Chattering problem in sliding mode control systems'. *Proceedings of the 2006 International Workshop on Variable Structure Systems, VSS'06*, pp. 346–350. IEEE-CSS, 2006.

135. V. Y. Utkin. 'Variable structure systems with sliding modes'. *IEEE Transactions on Automatic Control*, **22**(2):212–222, 1977.

136. V. Y. Utkin, J. Guldner, J. Shi. *Sliding Mode Control in Electromechanical Systems*. 1st edition. Taylor & Francis, London, 1999.

137. S. Valentinotti, B. Srinivasan, U. Holmberg, D. Bonvin, C. Cannizzaro, M. Rhiel, *et al*. 'Optimal operation of fed-batch fermentations via adaptive control of overflow metabolite'. *Control Engineering Practice*, **11**:665–674, 2003.

138. M. Velasco, O. Probst, S. Acevedo. 'Theory of wind-electric water pumping'. *Renewable Energy*, **29**:873–893, 2004.

139. S. R. Vosen, J. O. Keller. 'Hybrid energy storage systems for stand-alone electric power systems: optimization of system performance and cost through control strategies'. *International Journal of Hydrogen Energy*, **24**(12):1139–1156, 1999.

140. N. Vriezen, J. P. van Dijken, L. Läggström. 'Chapter Mammalian cell culture'. *Basic Biotechnology*, pp. 449–470, Cambridge University Press, 2001.

141. K. Walgama, S. Rönnbäck, J. Sternby. 'Generalization of conditioning technique for anti-windup compensators'. *IEE Proceedings on Control Theory and Applications*, **139**(2):109–118, 1992.

142. K. Walgama, J. Sternby. 'Conditioning technique for MIMO processes with input saturation'. *IEE Proceedings on Control Theory and Applications*, **140**:231–241, 1993.

143. Q. G. Wang. 'Decoupling with internal stability for unity output feedback systems'. *Automatica*, **28**(2):411–415, 1992.

144. Q. G. Wang. *Decoupling Control*. 1st edition. Springer, Berlin, Heidelberg, 2003.

145. M. Wei, Q. Wang, X. Cheng. 'Some new results for system decoupling and pole assignment problems'. *Automatica*, **46**:937–944, 2010.

146. S. R. Weller, G. C. Goodwin. 'Controller design for partial decoupling of linear multivariable systems'. *Proceedings of the 32nd Conference on Decision and Control*, pp. 833–834, San Antonio, 1993.

147. S. R. Weller, G. C. Goodwin. 'Controller design for partial decoupling of linear multivariable systems'. *International Journal of Control*, **63**(3): 535–556, 1996.

148. M. West. *Modelling and control of a sugar crushing station*. Master's thesis, Department of Electrical and Computer Engineering, The University of Newcastle, Australia, 1997.

149. J. Wolff, M. Buss. 'Invariance control design for constrained nonlinear systems'. *Proceedings of 16th IFAC World Congress*, Prague, Czech Republic, 2005.

150. W. Wolowich, P. Falb. 'Invariants and canonical forms under dynamic compensation'. *Siam Journal on Control and Optimization*, **14**:996–1008, 1976.

151. F. Wu, K. Grigoriadis. 'LPV based control of systems with amplitude and rate actuator saturation constraints'. *Proceedings of the American Control Conference*, pp. 3191–3195, San Diego, June 1999.

152. B. Xu, M. Jahic, S. O. Enfors. 'Modeling of overflow metabolism in batch and fed-batch cultures of *Escherichia coli*'. *Biotechnology Progress*, **15**:81–90, 1999.

153. Y.-S. Yang, Q.-G. Wang, L.-P. Wang. 'Decoupling control design via linear matrix inequalities'. *IEE Proceedings – Control Theory and Applications*, **152**:357–362, 2005.

154. D. C. Youla, H. A. Jabr, J. J. Bongiorno. 'Modern Wiener-Hopf design of optimal controllers, Part ii: The multivariable case'. *IEEE Transactions on Automatic Control*, **21**:319–380, 1976.

155. L. Zaccarian, A. Teel. 'The $\mathcal{L}_2$ (l_2) bumpless transfer problem for linear plants: its definition and solution'. *Automatica*, **41**:1273–1280, 2005.

156. L. Zaccarian, A. Teel. *Modern Anti-windup Synthesis: Control Augmentation for Actuator Saturation*. Princeton University Press, New Jersey, 2011.

157. G. Zames. 'Feedback and optimal sensitivity: model reference transformations, multiplicative seminorms and approximate inverse'. *IEEE Transactions on Automatic Control*, **26**:301–320, 1981.

158. H. Zhang, Y. Zheng, Q. Liu, X. Tao, W. Zheng, X. Ma, D. Wei. 'Development of a fed-batch process for the production of anticancer drug tat_m – survivin(t34a) in *Escherichia coli*'. *Biochemical Engineering Journal of Bioscience and Bioengineering*, **43**:163–168, 2009.
159. A. Zheng, M. V. Kothare, M. Morari. 'Anti-windup design for internal model control'. *International Journal of Control*, **60**(5):1015–1024, 1994.
160. K. Zheng, A. Lee, J. Bentsman, C. Taft. 'Steady-state bumpless transfer under controller uncertainty using the state/output feedback topology'. *IEEE Transactions on Control Systems Technology*, **14**:3–17, 2006.

Index